An introduction to rice-grain technology

An introduction to rice-grain technology

Kshirod R. Bhattacharya
and
Syed Zakiuddin Ali

WOODHEAD PUBLISHING INDIA PVT LTD

New Delhi

Published by Woodhead Publishing India Pvt. Ltd.
Woodhead Publishing India Pvt. Ltd.,
303, Vardaan House, 7/28, Ansari Road,
Daryaganj, New Delhi - 110002, India
www.woodheadpublishingindia.com

First published 2015, Woodhead Publishing India Pvt. Ltd.
© Woodhead Publishing India Pvt. Ltd., 2015

Woodhead Publishing India Pvt. Ltd. ISBN: 978-93-80308-58-6
Woodhead Publishing India Pvt. Ltd. e-ISBN: 978-93-80308-10-4

Typeset by Mind Box Solutions, New Delhi
Printed and bound by Replika Press Pvt. Ltd.

Contents

Preface

Rice is the most important staple food of human kind. A knowledge of the science and technology of the rice grain is thus important. A number of excellent books are now available on the subject. There is scope for many more.

I recently wrote a book entitled "Rice Quality: A Guide to Rice Properties and Analysis," published by Woodhead Publishing, Cambridge, in 2011. It is essentially an advanced treatise on the science of the rice grain. Now there is scope for writing a simple book on rice-grain technology, viz. methods and equipment for its drying, storage, milling, parboiling, etc. Books on this topic do exist. However, there is a huge pool of unmet demand for a *simple* book on the subject.

Rice/paddy is processed in enterprises that operate at various scales and levels of sophistication. These range from home- to cottage- to small- to very large-scale levels. As things stand today worldwide, the lion's share of the grain is handled in enterprises operating at the former levels, over 90% of world's rice being grown and processed in Asia.

To illustrate: In India (which accounts for about a fifth of the world's annual production of some 700 million tonnes of paddy), rice is processed in over 100,000 micro to macro enterprises. We can leave out the vast majority of these, which are really micro units consisting of perhaps a single hulling-cum-polishing machine each, and hence incapable of absorbing any modern knowledge. Even so there are some 20,000–50,000 small or large full-fledged multi-equipment rice mills dispersed throughout the country. At least half of these also run parboiling systems. The situation is not very different in the rest of the rice countries of Asia (south, southeast and east Asia). Whether small or middling, these enterprises run a fairly complex group of systems. But their operations are based largely on knowledge gathered by words of mouth, rules of thumb and the 'technology-suppliers' handouts; this despite early signs of a churning going on in the industry with wide opportunities arising and desire for quality and change.

These enterprises are starved of knowledge. There are no popular-level publications describing the basic theory and practice of rice technology presented in simple non-jargon language. This should give an idea of the vast potential demand. There is no doubt that a publication of this type will serve a felt need and will be widely welcome.

A second such potential target group is the community of students. Be it undergraduate or graduate, students who are exposed to Food Technology for the first time generally find it difficult to cope with scholarly presentations of the subject that are available. They too would benefit from being first offered a simple text book as an introduction to the subject, after which they no doubt should be encouraged to upgrade their knowledge by referring to more advanced books. And their number can be visualised from the fact that not only agricultural but also general universities these days have started offering courses in food science/technology throughout India.

I had previous experience of the potential too. The Government of India in the 1980s had set up a Rice Milling extension scheme, under which a branch had been set up in the Central Food Technological Research Institute (CFTRI) at Mysore in the late 1980s when I was the Head of the Department of Grain Science and Technology there. We brought out a series of simple booklets on various aspects of rice-grain technology. These booklets became immensely popular (and portions were also extensively copied verbatim in some publications, of course without acknowledgement!). But the curse of unpriced extension booklets is that they get quickly exhausted, going mostly to people who hardly need them and largely missing the target who really need them. So the effort was not as beneficial as it could be. But it taught me a lesson on the value of simple writing and the existence of a vast unmet hankering for information.

The present book is a culmination of that process of thinking.

Accordingly I have styled the book as an *introduction* to the subject. I have followed four general principles. First, I have kept the language as simple as possible, avoiding the use of jargons as much as I could. Second, I have consciously forgone the time-honoured practice followed in scientific presentations of authenticating each statement with the source (reference) or data. This otherwise essential practice would have been of no benefit to the untrained target audience and, on the contrary, would have made the text sombre and uninviting. Third, the attempt was largely to summarise only the up-to-date accepted knowledge in the topic. Fourth, at the same time, by adding a list of well-known books and texts as 'Further Reading', in which the topic has been discussed in detail, an avenue has been provided for the more inquisitive or advanced reader to explore deeper into the topic.

Let me confess something here. As a result of the procedure adopted as above, the presentation has often become an extended but simplified summary of the expert discussions in the 'Further Readings' mentioned above. These texts, therefore, have been extensively consulted and unabashedly used while preparing the present book, for which our indebtedness to these texts is acknowledged in advance.

I first intended to write the entire book on my own. But later I realised that the effort would be too much for me in view of my advancing age. So I roped in my first PhD student and later a valued colleague for decades, Dr. Syed Zakiuddin Ali, to write three Chapters 2, 5 and 10. I need hardly mention that he fulfilled his responsibilities with his usual competence. He also took upon himself the responsibilities, as a co-author for the project in various other ways, for all of which I am indebted to him.

I am indebted to Dr. B.V. Mehta, Executive Director of Solvent Extractors' Association of India, for kindly providing me with several reports, notes and information relating to rice bran and its oil. The help of Mr. R. Sathyanarayana, Mr. M.K. Suresha and Ms. K. Shobharani in preparing the typescript and the ready and helpful cooperation of the publishers are gratefully acknowledged.

Mysore   Kshirod R. Bhattacharya
April 2015

Dedication

Dedicated to the memory of the pioneers of India's rice-grain science and technology efforts: W.R. Aykroyd, B. Sanjiva Rao, A. Sreenivasan, V. Subrahmanyan, H.S.R. Desikachar and A.N. Bose.

1
On rice and rice country

This book is intended to be an introduction to rice-grain technology. The processes or the systems that are required to be applied to the rice grain after or from the time the grain is harvested in the field up to the time it reaches the consumer household is the subject matter of the proposed discussion.

However, an awareness of the background situation to start with may not be out of place here. Questions like what is rice, where does it grow, who grows it, who eats it, what is their lifestyle – these are not entirely irrelevant to the intended processes. The same is true of the genetics, botany, agronomy and trade of rice. An awareness of these issues as a background would help to provide perspective to the actions.

It is these background issues that are presented as a preliminary to the descriptions of the technologies to follow. Some thoughts on the geographical, historical and social background of rice are presented in this chapter. An outline of the agronomy and botany of the rice crop and its production and trade is presented in Chapter 2.

1.1 Geographical, historical and social background of rice

Growing of food grains, especially cereals, formed a watershed event in human history. Until then food had to be gathered or hunted every day or the every other day, for all other foods are perishable and cannot be stored for long. Cereals on the other hand are foods that are 'dry' and hence can be stored. As a result, cereals can be grown once (or twice) in a year but can provide food all the year round. Growing of cereals thus is one of the most significant events in the progress of human civilisation. Cereals became the staple food for humankind.

There are three major cereals, namely rice, wheat and maize. These are grown roughly in equal proportions – approximately 700 million tonnes of annual production of each. Apart from these, there are some minor or less important grains, such as sorghum, barley, various millets, and so on, whose total output may come to perhaps another 300–400 million tonnes. Rice is thus, though obviously important, as we shall see, only one among a few staple foods of humankind. Yet various historical, geographical and social backgrounds have combined to make the rice grain a unique food grain in human history. Many paradoxes and many peculiarities are associated with rice which contribute to the backdrop that makes rice unique. These paradoxes and peculiarities are proposed to be presented here. It is believed that an awareness of this background would provide a better perspective to understand the place or appropriateness of the technologies to be discussed later.

1.1.1   The puzzles associated with rice

The first thing to note in relation to the rice crop is a paradox regarding the growing of the crop. All agricultural scientists have said that the rice plant is very versatile and adaptable and can be grown widely. Indeed rice is grown up to 53° North latitude to 40° South, in some of the hottest parts to among the coldest regions of the world, from the mean sea level up to 3000 m above, in some very dry regions (negligible rainfall) to one metre under water. Yet the paradox is that 90% of world's rice crop is grown in a relatively tiny area of the earth in south, southeast and east of Asia (Fig. 1.1). Agricultural scientists have tried to explain this extraordinary concentration of rice cultivation in the following way: the area is characterised by vast expanses of flat land; several mighty rivers (Yangze Kiang, Hoang Ho, Red river, Mekong, Chao Phraya, Irrawaddy, Brahmaputra, Ganga, Godavari...) and their vast deltas; heavy and prolonged precipitation from the monsoon; and a very high population density (see below). It has been said that no other cereal crop would grow well under such heavy precipitation and water-saturated soil, combined with, for the most part, a warm and humid weather. As a matter of fact, the area is so well marked with a contiguous area, common rice culture and consumption, and many common cultural characteristics (see below), that it can be considered as a distinct region. We can call it the 'Rice Country'.

The second paradox associated with rice is the fact that most authorities are unanimous in claiming that rice feeds half the world. Any number of statements can be quoted from authoritative accounts on this score. At any rate, whether half the world is a fact or perception, there is no doubt that rice is the staple food of the largest number of human population of the world. The paradox is obvious. As mentioned above, rice is only one of several cereal crops. No doubt it is a huge crop, but it is not the only cereal crop, nor it is by far the largest one. If one goes by the production figures, output of rice comes to approximately a little less than 30% of the world's cereals output. Then how does it feed half the world, or at any rate the largest number of people of the earth, is a mystery. Actually, we can observe that a part of the answer perhaps lies in the extraordinarily high population density in the Rice Country, as shown below. The other part, we can surmise, is apparently the fact that, worldwide, an appreciable to a large proportion of all other cereals output is used as animal feed or for production of starch and other industrial products; but not rice. Rice is too precious, especially in the context of the Rice Country. Every grain of rice here is used as human food. If any rice is used by the industry, it is only for production of other kinds of rice food.

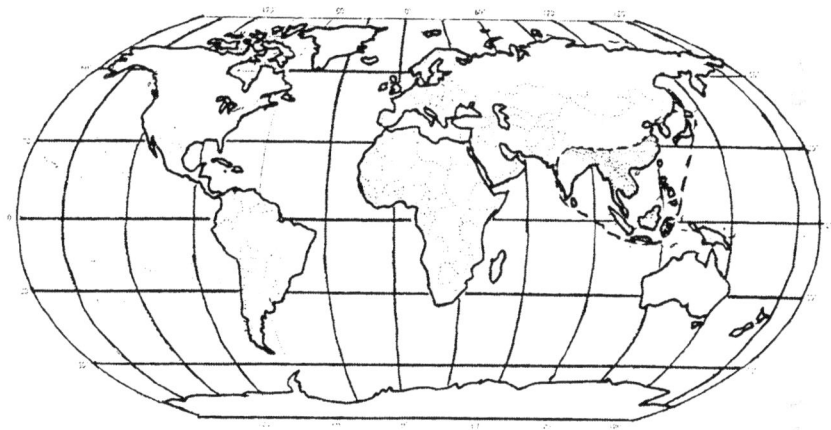

Figure 1.1 World map, highlighting the Rice Country
[Reproduced with permission from Bhattacharya (2011)].

In this connection, another crucial feature of the Rice Country, and therefore indirectly of rice, is the statistics related to land area, arable area and the population of the countries within the region. Some of the extraordinary data are shown in summary form in Table 1.1. The full data show that all the countries within the region are characterised by two conspicuous features. First, a high proportion of their land area has been converted into arable land (mean, 21%). This figure is nearly double the proportion (mean, 11%), so converted in the rest of the world. Second, these countries are all characterised by an extraordinarily high population density. In aggregate, we can see that the Rice Country is characterised by only 14% of the world's land area, 25% of the world's arable land area, but a staggering 54% of the world's human population. The other way of looking at it is to note that the rest of the world with 86% of the world's land area and 75% of world's arable land area carries a population load of only 46% of the world. Such is the extraordinary population pressure in the Rice Country and, equally, such is the ability of rice to sustain this population.

Table 1.1 Land area and population density in Rice Country

Country	Land area (% of world)	Arable area		Population (% of world)	Population Density (persons/sq km)
		(% of its land area)	(% of world arable area)		
Rice Country	14	21	25	54	202
World excluding Rice Country	86	11	75	46	28

Source: Table 1.1 is a summary of detailed data presented in Tables 1.2 and 1.3 in Bhattacharya (2011) for approximate period 2005–2010. Original data sourced from US Government sources cited in Bhattacharya (2011).

Another characteristic feature of the rice country is that all the countries within the region (barring Japan, which preceded the others in industrial development by half to three-fourths of a century or so) are currently in the developing stage. Most of them were flourishing civilisations roughly two to three centuries ago, missed the industrial revolution, became colonies of (or were dominated by) the newly industrialised European countries, sank into abject poverty, and are now showing signs of reversing the trend and standing up again.

The next characteristic feature of rice is the fact that rice is not simply a food but also a livelihood. This can be understood in the socio-economic and historical contexts just mentioned. Once the countries fell into abject poverty, rice became not only a food but a prime livelihood. If one considers the gigantic number of people involved in small-scale rice cultivation, its handling and processing, its milling, making products out of it and its trading, one can realise why it is said that rice provides the largest economic activity on earth. Indeed, the international rice data bank, the Rice Almanac (Maclean et al. 2002), makes the following statements among many others of a similar kind:

- Domestication of rice ranks as one of the most important developments in history. Rice has fed more people over a longer period than any other crop.

- Rice is the staple food for the largest number of people on earth.
- Rice is eaten by nearly half the world's population.
- Rice farming is the largest single use of land for producing food.
- Rice is the most important economic activity on earth.
- Rice is the single most important source of employment and income for rural people.

Finally, the last characteristic feature to be noted about rice and the Rice Country is that rice is not just a food and livelihood but also a culture. It is a part of community life, social organisation and culture. In most languages of the Rice Country, the words for rice and food are synonymous. When visitors come home, they are invariably offered rice, meaning food. All the activities of the rural population revolve round the seasonal activities related to rice production. All cultural activities and life styles are modulated with the progress of the season and activities related to the production of rice, including ploughing, land preparation, sowing, transplanting, weeding, harvesting and so on. Each phase is connected to a particular season, rituals, celebration, community activity and religious events.

It is also pertinent to note that in the countries within the region, rice and everything connected with rice are considered to be endowed with spirits that require our protection, respect and propitiation. The seed is not just seed but a container of the spirit of rice which needs to be protected and propitiated to result in a bounty of harvest. Every aspect of rice production is associated with proper rituals, hoping to be rewarded with a bountiful harvest and happy community life. It is extraordinary to note that these rituals and customs are comparatively speaking quite common in different countries of the region despite differences in the popular religions (Islam in Bangladesh and Indonesia, Christianity in Philippines, Buddhism in Thailand, Sri Lanka and Myanmar, Shintoism in Japan, Confucianism in China, and Hinduism in India). What is astounding is that this great attachment to the perceived sacredness and preciousness of rice is true even in Japan which has been a highly industrialised country for at least half a century. As a matter of fact, according to Japanese folklore, the Japanese are nothing but rice farmers on an island. That is, the Japanese are thought to be characterised by insularity and rice cultivation or, to put it differently, settled life and a plot of land. Extending from this, according to the Japanese ethics, the reward of one's life is to sustain a paddy with all one's might. The ethical core of the Japanese belief is *issho-kenmei*, which stands for 'do your best', and literally means 'maintain a piece of land for all your worth'. So strong is the attachment of Japanese to rice. In other words, there is something characteristic of rice and of the region of the Rice Country which makes rice not just a food or a livelihood but something much more precious.

To summarise, we can say that the Rice Country is characterised by the following specific features:

- The area comprises the south, southeast and east of Asia. It is a contiguous region, comprising of a total of 20–25 huge (China) to tiny (Singapore) countries (Fig. 1.2), whose primary staple crop is rice. Over 90% of world's rice is grown and consumed here.

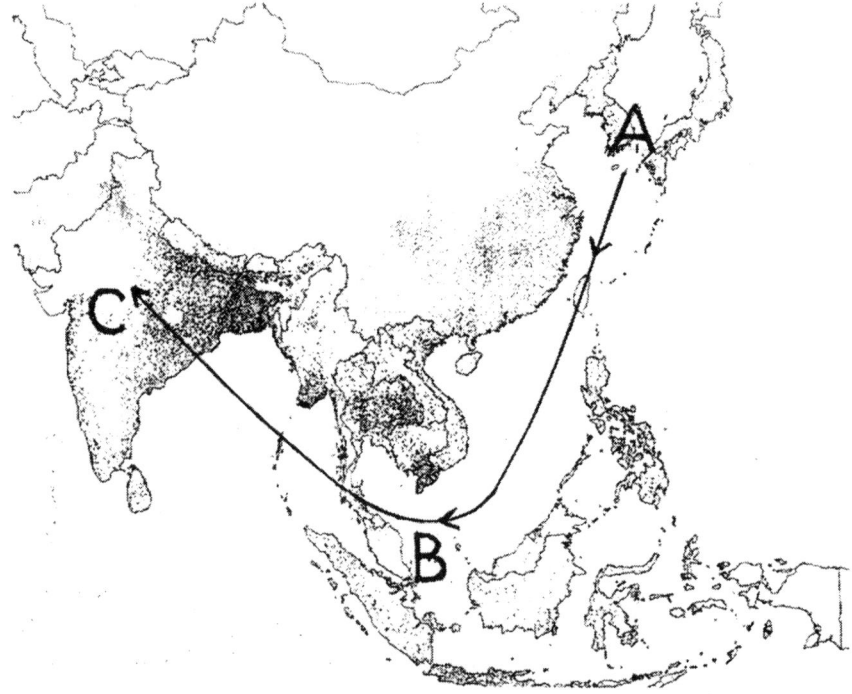

Figure 1.2 Map of Rice Country. Over 90% of the world's rice is grown and consumed in 20–25 tiny to giant countries in east (zone A), southeast (zone B) and south (zone C) Asia. The original map is reproduced from IRRI (2007). The arrow shows the gradation of texture of cooked rice from soft and sticky (zone A) to firm and free-flowing (zone C).

- With small exceptions, the area is under the common influence of heavy precipitation from the monsoon. By and large all the major countries in the area also have huge flat lands and a relatively large proportion (21%) of their land has been converted into arable land as compared to the proportion (11%) in the rest of the world.

- The region is blessed with several mighty rivers and their huge deltas.
- All the countries have an extraordinarily high population density (on an average, 202 persons per sq km) as compared to the rest of the world (28 persons per sq km).
- Despite ethnic, linguistic and religious differences, the entire area has adopted fairly common cultural symbols, rituals and myths about rice.
- The entire area was once a highly civilised region, but came under colonial domination and subsequent abject poverty, and is now showing signs of rising again.

With such extraordinary commonality, the entire area deserves to be considered and recognised as a common and distinct region in the Earth.

1.1.2   Diversity within unity

With all these commonality in agriculture, food habits, life style and cultural traits, it would be wrong to say that the Rice Country region is an undifferentiated homogenous mass. There is actually a unity with some underlying systematic diversity. In fact it is a case of one common region but having three types of rice in three fairly distinct zones. The three zones are: (A) east and northeast Asia, (B) southeast Asia and (C) south Asia (Fig. 1.2). There is a very interesting diversity as well as a gradation of the type of rice among these three zones which can be summarised as follows.

- Two of the zones (zones B and C) lie in the tropical region and partly in the subtropical region. This is the major part. A smaller part (zone A) falls in the temperate region (Japan, Korea and northern and eastern China) (Fig. 1. 2).
- Correspondingly, the prevailing agricultural growing temperature varies accordingly. Thus rice is grown in extreme cold conditions in the northernmost part of Japan (Hokkaido island) in zone A, as against that grown in the warm and humid climate of south and southeast Asia (zones B and C).
- There is a striking gradation in the type of rice grown among the three zones. The rice grown in the relatively cold and temperate climate of east and northeast Asia (zone A) is called the japonica type of rice (refer Chapter 2). The grains are relatively short in size and roundish in shape. They also have a relatively low content of

amylose starch (< 20%) and cook rather soft, moist and sticky. In contrast, the rice grown in the relatively warm, humid, tropical and subtropical areas of southeast (zone B) and south (zone C) Asia is different. It is indica type in both the zones, the grains of which are usually rather longish in size and slender in shape. But there is a difference between the two latter zones (B and C) also. Though both grow indica rice, the rice in zone B has by and large an intermediate content of amylose starch (20–25%) but that in zone C possesses a high content of amylose starch (> 25%). So rice in zone B by and large cooks intermediate but that in zone C generally cooks relatively hard (firm), dry and discrete.

• With the above difference in the type of rice grown, there is a very interesting gradation in the texture of corresponding cooked rice, as shown in Fig. 1.2. While in the east and northeast region (zone A), the rice by and large cooks soft, sticky and moist, that in south Asia (zone C) cooks relatively hard, integral and dry. Indeed people in south Asia are so much enamoured of relatively hard-cooking rice (high amylose) that a vast section of people here (about 65% of rice eaters) prefer to eat parboiled rather than raw (i.e. non-parboiled) rice (refer Chapter 7). One should note that parboiling makes rice still more hard-cooking. The rice of Southeast Asia (zone B), located in the middle, is indeed intermediate between these two extremes. So there is a gradual and systematic gradation of the preferred rice texture from east and northeast to southeast to south of Asia (Fig. 1.2) – soft (zone A) to intermediate (zone B) to hard (zone C).

• As a parallel to the above differences in the type of rice, there is another most interesting and amusing difference among the respective people. This is in relation to their liking for the process of ageing of rice. It is well known that rice ages after harvest and progressively becomes hard-cooking with time of storage (refer Chapter 6). While people in south Asia (zone C) heartily dislike soft and sticky-cooking, freshly harvested rice and prefer aged rice, those in Japan and other regions of northeast Asia (zone A) prefer the opposite. They like freshly harvested rice and heartily dislike aged rice. People of Southeast Asia (zone B), true to their middle position, are indeed in between. On the whole they prefer aged rice but are not so emotionally attached to it.

• Another interesting difference exists between the three zones. While a majority of rice snack products made in east (zone A) and southeast (zone B) Asia are a variety of wet-ground and cooked rice cakes, the most common rice snack in south Asia (zone C) is granular whole-grain products (refer Chapter 8).

- Finally, waxy rice (which has no or negligible amylose starch, so cooks very sticky and soft) is used fairly commonly, both for cooked table rice and for snacks, in southeastern and northeast (Asia zones A and B). But waxy rice is not at all common in south Asia (zone C).

So the Rice Country is a paradox within a paradox: a striking unity containing a systematic diversity within.

Further reading

The theme of this Chapter has been further developed, with more data, in:

- Bhattacharya (2011)

2

Agronomy, production and trade of rice

Rice is one of the three major food crops of the world, others being wheat and maize. It forms the staple diet of more than half of the world's population. The global production of rice, in terms of paddy, is estimated to be 745 million in tonnes in 2014, grown over 163 million hectares (FAO, 2014). As explained in Chapter 1, although it is grown under diverse agro-climatic conditions around the globe, it is the major crop of countries in the south, southeast and northeast Asia, which account for more than 90% of the world's produce. It plays a crucial nutritional and cultural role in societies throughout the countries of this region. It is also of immense importance to food security of these countries. The top two rice-producing countries, China and India, together contribute to more than half of the world's produce.

2.1   Origin and domestication of rice

Rice, as we now know, is a cultivated annual crop, but its ancestors existed as perennial plant in the wild habitat. It is most likely that the prehistoric people in the above-mentioned geographical areas gathered this rice as a food supplement to game and fish they hunted. The gathering of wild-growing perennial rice preceded the beginning of agriculture in the humid tropics of Asia, probably around 8,000–13,000 B.C. It is not clear as to when the annual forms appeared from their ancestral wild perennial types, but domestication might have been possible only after the annual forms were identified.

Presence of paddy grains, or parts of the plant, in the archaeological finds of ancient civilisations in more than 100 sites in northern and eastern India, northern Southeast Asia, southwest China, and more recently in central, as well as in northeast China, have thrown much light on the origin, spread, domestication and cultivation of rice through the history. Much advancement has taken place on the understanding of both archaeology (archaeobotany) and rice genetics (genome sequencing and genetic markers) in the past couple of decades that has helped in a better understanding and placement of archaeological finds in a reasonable time frame in the history. It is now postulated (Fuller 2011) that the rice plant that was first domesticated and cultivated was of *Oryza sativa* species, which is the prevalent species being cultivated even today. It is suggested that the domestication of the crop followed two independent pathways, one in China and the other in India. The ancestors of *Oryza sativa* are the *Oryza rufipogon*, (a predominantly perennial wild rice) and *Oryza nivara* (an annual wild rice), both native to the south and Southeast Asia. The earliest domestication traits have been found in the rice samples excavated at the archaeological site in the lower basin of river Yangtze in China and date back to around 6,500 B.C. On the basis of grain morphology, this domesticated rice is classified as belonging to the subspecies japonica, also known as *sinica* or *keng*.

Domestication of indica type, the other major subspecies, is suggested to have taken place through a separate domestication path, in the Gangetic plains in eastern India, about two millennia later, around 4,500 B.C. It is postulated that this happened upon natural hybridisation of the then existing semi-domesticated or wild 'proto-indica' rice with the fully 'domesticated' japonica. The japonica thereby became the donor of many domestication genes in indica. This postulation is based primarily on the reconciliation of genetic information on the main domestication gene, *sh4*. This gene is responsible for the reduction of grain shattering from wild to cultivated rice, and is believed to have originated only once, in japonica. The gene flow dynamics indicate its transfer to indica as a result of natural hybridisation.

The domesticated rice underwent many morphological and physiological changes when it was widely disseminated by cultivators into divergent geographical and ecological systems in humid tropics and subtropical zones, as the human settlements moved. Further dispersal, hybridisation and selection led it into temperate zones too, as for north as 53° latitude North. The varieties that were domesticated in the tropical islands of Indonesia, Java and around, near the equator, evolved into a separate distinct group that had different plant and grain characters than those of indica or japonica. Though grown in a limited geographical zone of these islands, these varieties have been recognised as a separate group and classified as Javanica. Some of these varieties have awns at the tip of the grain (*bulu*), while others are awnless (*gundil*). The classification of rice will be further discussed later. The combined forces of natural selection and cultural practices have added to the great ecological diversity exhibited by Oryza sativa cultivars. Over the millennia, the frequency of cross-pollination also declined so that the crop became more inbred than its wild ancestors.

The other cultivated rice species, *Oryza glaberrima*, was evolved in West Africa from its wild ancestor *Oryza barthii*, in the swampy basin of the upper Niger River. The primary centre of its domestication was probably formed in this region as late as 1,500 B.C. Two more secondary centres seem to have formed 500–700 years later, in the southwest regions, near the Guinean coast.

It is worth noting, however, that samples of *Oryza glaberrima* collected recently a few decades ago, from many parts of West Africa, show that it has lost some of its distinctive features, partly because of mixed planting with *Oryza barthii* and *Oryza sativa* in the same field.

2.2   Botany of rice

2.2.1   Classification

Rice belongs to the genus *Oryza* and the tribe *Oryzeae* of the family *Gramineae (Poaceae)*, i.e., the grass family. The genus *Oryza* contains 22–25 (still debated) recognised species, of which only two, viz. *Oryza sativa* and *Oryza glaberrima* are cultivated, and the rest are wild. *Oryza sativa* is grown worldwide, whereas *Oryza glaberrima* is grown only in the West African countries. However, *O. sativa* and *glaberrima-sativa* hybrids are now replacing *O. glaberrima* in many parts of Africa due to higher yields.

The *Oryza sativa* species was thought to be comprising of three subspecies, viz., indica, japonica and javanica, mentioned above, which evolved as a result of centuries of selection by man and nature for desired quality and adaptation to different ecological systems. The indica varieties dominate the tropical countries of south and Southeast Asia. The japonica varieties are prevalent in subtropical and temperate regions in Japan, Korea, northern China, Egypt, Italy, and Spain, while the javanica varieties are grown near the equator, in the islands of Indonesia. These subspecies could be distinguished on a few key plant and grain characteristics, such as glume size, number of secondary panicle branches (rachii), panicle thickness, grain size and shape, composition of the starch, and the texture of rice when cooked.

The indica varieties are tall, leafy, grow vigorously, produce a large number of shoots (tillers), and need a long duration to mature. They normally lodge (fall down) when more fertiliser is applied, but can grow fairly well even under poor soil conditions. The grain yields are rather low. They are also photoperiod sensitive. That means, the flowering, the grain formation and also some other phases of plant growth are initiated only when a specific day length period arrives. So, they need to be grown is specific season, or the time of the year. In order to overcome these drawbacks, breeders have developed semi-dwarf varieties which do not lodge, and also are photoperiod insensitive, meaning they could be grown in any time of the season. The rice cultivars from indica are characterised by having higher amylose content (24% to 30%) in starch. The grains normally have long to medium length, and are thin. The cooked rice has a non-sticky, fluffy and firm texture.

The japonica varieties are grown in subtropical and temperate regions. Their plants have short and sturdy straw; they do not fall down easily even when high fertiliser doses are applied (non-lodging) and exhibit high response to nitrogen fertiliser. It means they give more yields with more fertiliser. They are photoperiod insensitive and hence are not season bound. They mature early (short growing time). The rice has medium to low amylose content in the starch (18% to 23%). The grains are short and roundish, and the rice cooked from them has a sticky texture.

The javanica varieties, grown in the tropical islands of Indonesia, have large plant height with tick straw and have grains that are large and bold. They have medium amylose content and the texture of cooked rice is soft.

A new system of classification of rice varieties has been proposed by Glazsmann in 1987. This is based on the grouping of isozyme markers analysed from 1688 rice cultivars. The isozymes were formed into six groups. There were two major groups, which included the indica (group 1) and japonica (group 6) cultivars, and four small intermediate groups. The japonica group included the typical short- and medium-grained cultivars from temperate regions, as well as the tropical types that included the upland and javanica cultivars with a range of grain sizes and shapes. Groups 2 to 4 were minor groups consisting of cultivars from eastern India and Bangladesh. Group 5 included Basmati and other aromatic cultivars and premium quality rice. The isozyme study showed that the temperate japonica and the tropical javanica rice were indistinguishable, and they formed a continuum based on morphological characters. The javanica cultivars are therefore proposed to be considered as tropical japonica as per this classification.

2.2.2   Breeding for higher yields

Rice has been grown for centuries traditionally by relying on monsoon. However, there is a limitation in the yield of rice/paddy in this system. As population in countries depending on rice as the staple continues to grow, there is an increasing demand for higher production of rice. Attempts have therefore been made to meet this challenge. Establishment of the International Rice Research Institute in Philippines in 1960 was a major step towards this objective. It was realised that the two things that are necessary to increase the production of rice are: one, to increase the yield of rice per unit area of cultivation per crop; and two, to grow more than one crop per year. In order to achieve these objectives, the following three factors, considered necessary, need to be addressed.

First, the yield of rice is related to the shape and structure of the plant. Traditional varieties in tropical Asia grow vigorously, become tall and produce many shoots and leaves. The large number of long and drooping leaves shade each other, limiting the sunlight falling on the leaf blades. Bright sunlight is necessary to produce more food as it increases photosynthesis for formation and filling of grains by the plant. In addition, tall plants easily fall down (lodge) if more fertiliser is applied. Ideally, the plant should be short and stiff with moderate shoots and erect leaves. This helps in tolerating more fertiliser without lodging, thus increasing the yield. Also, erect leaves do not shade each other thus allow more sunlight to fall on leaf-blades to produce more food.

Second, dormancy, which is an important character of all seeds, limits and determines the number of crops that can be grown in a year. When seeds are placed in soil with water, they germinate and produce the plant. However, it is found in nature that many seeds do not germinate soon after harvest, but are able to do so after a lapse of certain period. This phenomenon is termed as dormancy. The seeds regain the capacity to germinate after their dormancy period. This is a natural mechanism to preserve life. If there is no dormancy, the seeds would start germinating when there is a rain during harvesting time or soon after, and so would go waste. Dormancy preserves the seeds, and enables them to germinate only in the next season or so, thus perpetuates the generation. The period of dormancy varies among different plants and varieties. Traditional rice varieties generally have long dormancy, while some modern varieties have practically none. If the dormancy is too long, it prevents the seeds to be used for sowing for a second crop soon after harvest. On account of this, the traditional varieties can be grown only once in a year. Only varieties with short dormancy can be grown twice, or even thrice, a year if irrigation is provided. Total absence of dormancy, on the other hand, is also not a desirable property, as mentioned above. A dormancy period of about 15 days is considered optimum for rice. This is one of the objectives in the breeding programmes for crop improvement.

The third point is the photoperiod responsiveness of the rice varieties discussed earlier. Varieties which are photoperiod sensitive can be grown only once in a year. If more than one crop in a year is desired, the varieties must be insensitive to photoperiod.

It is possible therefore to achieve higher yields and higher production of rice by addressing the above factors and incorporating appropriate methodologies in rice breeding programmes. However, apart from achieving the desired plant characters, other inputs are also necessary to see that the crop responds. It needs artificial fertilisers to produce more grains. Further, to get two or three crops in a year, irrigation facilities are a must, as rice crop needs plenty of water. A negative aspect of continuous cropping is the insect and disease build-up, to which the crop would become prone to. In order to prevent and control the same, use of some chemical pesticides then becomes essential.

The first task the International Rice Research Institute took up was to change the plant structure. They obtained a short variety, called "Dee-Geo-Woo-Gen" from Taiwan and crossed it with various tropical varieties to produce the first miracle rice, IR 8. This was high yielding, dwarf in structure, short in dormancy, and photoperiod insensitive. It produced remarkably high yields of paddy with nitrogen fertiliser application, and matured in one to two months earlier than the traditional varieties.

In the mid-60s, most of Asia was experiencing drought and potential famine conditions. So, IRRI decided to get IR8 out quickly to the rest of the world. And they did. It quickly became popular across the rice-producing countries, heralding the 'green revolution'.

The rapid spread of IR8 into many countries in tropical Asia, Latin America and West Africa was phenomenal. By the end of 1966, IR8 had been introduced to and adopted by about 60 countries in various geographic areas. However, IR8 was not perfect. Certain pests and diseases attacked IR8 easily. It had a high breakage rate during milling. The grain it produced was actually bold and chalky, so it didn't look as good to the consumer as the traditional polished rice. In addition, it had a high amylose content, which meant that it hardened after cooking, not a desirable property in some countries. Many more varieties have, therefore, been bred since then to address these drawbacks by crossing short varieties with various tall varieties, and to include other desirable characteristics such as resistance to diseases and insects, adaptability to various ecological conditions and a desired cooking quality. Another semi-dwarf variety, IR36, that had many desirable characteristics, developed later, proved very successful and became very popular. By 1980, at least 11 million hectares were planted with IR36 around the world, and the area under all the high-yielding varieties in the world reached 40% of the total rice. Further on, IR72, released in 1990 out-produced even IR36. A large number of semi-dwarf varieties in different countries of tropical Asia have been developed in due course of time, either using the semi-dwarf gene in TN1 or IR8 or selecting crosses made by IRRI.

Thus was born the high-yielding technology in rice cultivation, which greatly increased the rice output. The world average of paddy yield, the product of thousands of years of experience, was about 2.0 tonnes per hectare in 1965. However, 35 years later, as the Green Revolution was introduced and spread, it doubled, reaching 4.0 tonnes per hectare in 2000. It is estimated that about 80% of the total area under rice is now covered by high-yielding varieties globally. The rice varieties and technologies developed during the Green Revolution have increased yields in some areas to between 6.0 and 10 tonnes per hectare.

2.2.3   Hybrid rice and super hybrid rice

Hybrids are obtained by crossing two parents (say two rice varieties) which are genetically different from each other. The offspring of such a hybrid has more vigour than the parents. In case of rice, it implies more yield, of about 20–30% higher as compared to the parents. However, this vigour disappears after the first generation (called F1). So farmers, therefore, cannot use the seeds obtained from a hybrid crop. They have to use fresh stock of F1 seeds every time they grow the crop, to reap the benefit of higher yields.

Rice is mainly a self-pollinated crop. Each rice plant produces its own pollen which fertilises the ovary and produces the seed, i.e. the rice grain. On account of this, rice has been a poor candidate for commercial hybridisation because one would have to find a way to produce the F1 seeds continuously for growing the hybrid crop. In 1970, Chinese researchers came out with a technology to produce large quantities of F1 seeds to supply to the farmers. The system developed has been successfully implemented and is widely practice in China. For this system to work, one needs three different lines of plants. First is a male sterile line, to ensure no self-pollination takes place (fertilisation is by another variety planted together). Second is the maintainer line, to produce and maintain the male sterility, and the third, a normal parent, a restorer to give the F1 hybrid.

Large-scale production of hybrid rice in China was started in 1976. By 1991, more than 50% rice area in China was covered with hybrid rice. The trend continues to be so even today. In order to boost the rice yields still further, China has embarked on a 'Super High-Yielding Hybrid Rice', or the 'Super Rice' Programme. A recently developed super hybrid rice has given a yield of 13.9 tonnes per hectare in 2011, in large-scale demonstrations.

Chinese success with hybrid rice encouraged IRRI, and many countries, like India, Vietnam, Philippines, USA, Bangladesh, Myanmar, and Indonesia to embark on hybrid rice programme. Although hybrid rice has been a great success in China, it has not been the case in other countries. Initially, Chinese hybrid seeds were introduced in these countries, but it met with many bottlenecks. The main constraints are: poor grain quality (higher content of chalky grains), poor cooking quality, reduced milling yield, higher content of broken rice, and susceptibility to many diseases like blast, bacterial leaf blight, bacterial leaf streak and some pests, when grown under different agro-climatic conditions. High cost of the Chinese seeds and perennial dependency on it, were also a deterrent. Many countries, therefore, started their own research for the development of hybrid rice to suit to their agro-climatic and domestic quality requirements. A 'Hybrid Rice Development Consortium' (HRDC) has been formed in 2008 to help the R&D and the exchange of knowledge towards the development of appropriate hybrids. In the year 2011, the membership of the HRDC stood at 57 organisations from 22 countries. Majority were from private sector. India had the largest group of 17 organisations from private sector, while 3 were from the government.

There is growing concern, however, regarding the increasing interest and investment of private sector in the business of hybrid rice seeds in some countries, as it would lead to placing a county's food security at the hands and mercy of private sector corporates. Farmers who plant hybrid rice have to return to the company every cropping to buy new seed. The small farmers, who form the major proportion of rice farmers in the Asian rice-producing countries, are expected to be worst hit. According to these concerns, the corporate investors are trying to take control of the world's rice farms and the world's rice supply, from the seed up. It is felt that the involvement of private investors is not about the performance or increase in rice production, but the control it offers over farming. For hybrid rice, therefore, many more challenges are to be met to make further headway.

2.2.4   Rice plant morphology and pollination

Rice plant is a typical grass. Its morphological features are shown in Fig. 2.1. The plant has a fibrous root system with erect jointed stems called culms that develop long and flat leaves along its length. It has a semi-aquatic lifestyle, requiring water particularly during the reproductive growth phase. It develops multiple tillers (shoots), each consisting of a culm and leaves, with or without a panicle. The panicle emerges on the uppermost node of a culm, from within a flag-leaf sheath and bears the flowers in spikelets. The culm consists of a number of nodes and hollow internodes that elongate as the plant grows. Single leaves develop alternately on the culm, consisting of a sheath which surrounds the culm, and a long, flat leaf blade. The leaf base forms a collar known as junctura between the sheath and the blade.

The panicle (Fig. 2.2), which carries the bunch of flowers, emerges from the sheath of the flag-leaf (the topmost leaf). It consists of a primary or a main axis called primary rachis, branching into secondary tertiary ones, which bear still smaller branches, known as rachilla. Each rachilla bears a spikelet at its tip, which contains a single floret and two glumes. The floret is enclosed by a rigid, keeled glume lemma, which is sometimes extended to form an awn. It envelops partially the smaller glume palea. The lemma and palea form the two halves of the grain husk. The floral organs are protected within lemma and palea. Each floret contains six stamens and a single ovary with two branches of stigma on its top.

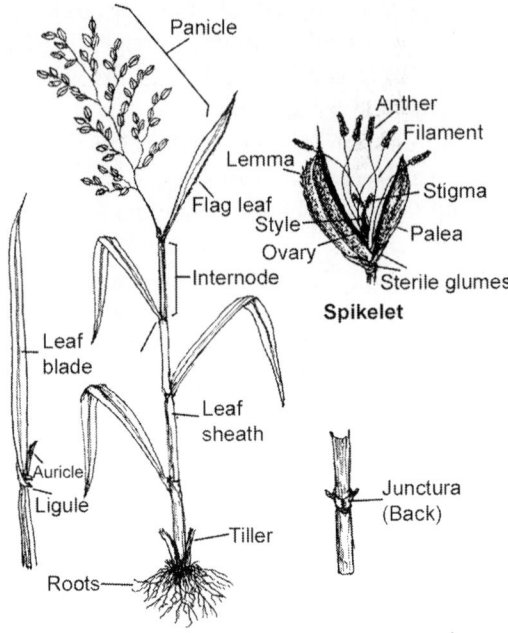

Figure 2.1 Paddy plant and its parts

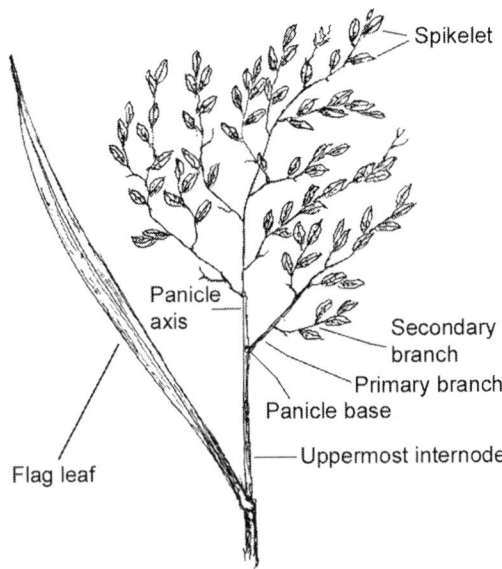

Figure 2.2 Paddy panicle and its parts

Pollination

When the flower is ready to bloom (anthesis), lemma and palea open and the stamens (the male parts consisting of anther and filament) elongate rapidly and emerge. The stigma, the female part, is exposed as well. The anthers burst and shed the pollen (male cells) on stigma. This process is known as pollination. After some time, the two glumes come back to their original position and get tightly interlocked. The whole operation may take 2–4 hours. Generally, pollination in rice takes place between about 10 am and 1 pm depending on the weather conditions. Rice is mainly a self-pollinated crop. That means the pollen grains pollinate the stigma of the same plant. The male and female cells brought together in this manner get fertilised and start growing inside the glumes to become the grain.

The difference in the blooming of the top and bottom spikelets in a paddy panicle varies from 3 to 7 days, and the maturity between the top and bottom grains varies from 8 to 12 days. The difference in the maturity time among the grains in a panicle, and between different panicles in a plant, is the basic cause for the harvesting problem in rice. In shorter panicles, grains have more uniform maturity than in longer ones. Uniform maturity minimises cracking and shedding losses. The number of grains in a panicle normally varies from 75 to 150. A short, stiff, strong and erect panicle can bear a large number of grains without drooping down.

2.2.5   The rice grain

The rice grain is a ripened ovary, with lemma, palea, rachilla and the sterile glumes firmly attached to it. It is in fact a dry fruit containing a single seed. Other parts of the fruit are not well developed and therefore become remnant. The structure of the paddy grain is shown in Fig. 2.3. The outermost covering of the grain, the husk, consists of lemma and palea, which is removed during shelling operation while milling. The unpolished rice within the husk is called brown rice because of the brownish colour of its outermost layer, the fruit wall or the pericarp. Next to it lies the seed or coat or the tegmen. Before fertilisation, it contains highly specialised tissues called ovules, which develop into embryo and endosperm after fertilisation. The highly compressed outer layers of ovary become the seed coat. A single seeded fruit in which the pericarp and seed coat are fused is called 'caryopsis'. All cereal grains are therefore botanically called as caryopsis.

Next to the fused layer of pericarp and seed coat lies the aleurone layer which forms the outermost part of the endosperm. Aleurone layer contains high amounts of protein, fat vitamins and minerals. The germ (embryo) lies on one side of the grain next to lemma. The germ contains the young shoot or stem portion of the future plant, called plumule, and the root portion, called radicle. Both plumule and radicle are enclosed by two cap-like structures called coleoptile and coleorhiza, respectively. The plumule, radicle and their associated structures are collectively called embryonic axis. These are bound on their inner side by scutellum. Scutellum is the remnant of a cotyledon. Cereals are called monocots because the scutellum represents a single cotyledon. Embryo and scutellum are rich in fat, proteins and vitamins. When rice is milled, some part of the scutellum is retained in the milled rice. More the scutellum is retained; more is the nutritive value of milled rice.

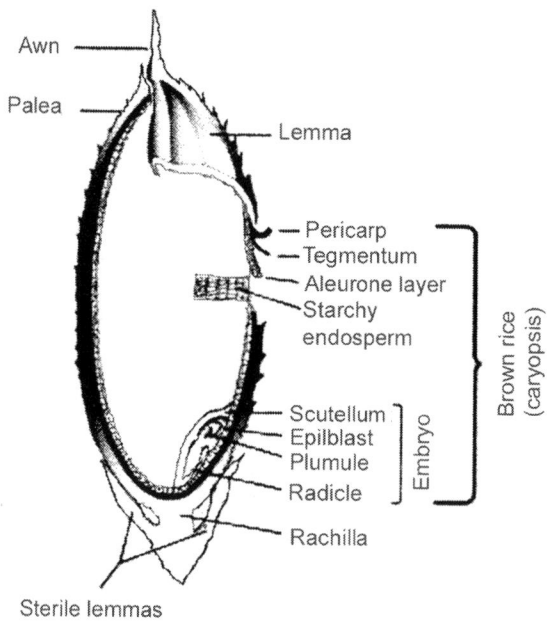

Figure 2.3 Structure of rice grain
[*Source:* Mclean et al., 2002, used with permission from IRRI]

The weight distribution of various anatomical parts of brown rice is: pericarp – 1% to 2%, aleurone and seed coat – 4% to 6%, germ (embryo) – 2% to 5%, and endosperm – 89% to 94%.

It is estimated that the approximate aggregate number of endosperm cells in a rice grain is about 180,000. These cells are of three kinds: plate like, long tube like, and short tube like. The arrangement of cells is like bricks in the wall along the longitudinal axis. This makes the grain susceptible to cracking across the length along the cell wall line.

Lot of variation exists in the size and shape of rice grain, as well as in grain weight. In dimension, a milled rice grain varies from 4 to 7.7 mm in length, 1.7 to 2.9 mm in breadth, and 1.4 to 2.0 mm in thickness. The length to breadth ratio varies from 1.6 to 3.8. The grain weight varies from 7 to 30 mg.

2.3   Growth and development of rice plant

Starting as a germinating seed, growing into a plant, reaching a stage of maturity, and drying up after producing seeds for the next progeny, the life span of the rice plant may spread over a period of 100–210 days. Majority of rice varieties, however, have a life cycle of 110–150 days from sowing to the harvest. Temperature and day length are the two environmental factors that affect the development of rice plant. Usually, the photoperiod sensitive, traditional tall varieties have a life cycle of more than 150 days, while the modern, photoperiod insensitive, short and dwarf varieties have a range of 100–130 days.

Rice plant growth can be divided into three agronomic phases of development: the vegetative phase – from seed germination to panicle initiation, the reproductive phase – from panicle initiation to flowering/anthesis/heading, and the ripening phase – from flowering/anthesis to full maturity. These main phases overlap each other within a rice hill or a rice crop. The ripening phase physiologically does not start until 3 weeks after fertilisation. Under tropical conditions, the duration of the reproductive phase (about 35 days) and the ripening phase (about 25 to 35 days) are fairly constant for the modern dwarf, as well as the traditional tall varieties. The difference lies in the duration of the vegetative phase, whose length may be inherent or dependant on the sensitivity of the cultivar to day length and temperature. Low temperature affects the vegetative and ripening phases. IR8, with a growth period of 120 days in tropics, may take 180–200 days in temperate areas, or at high altitudes in tropics. Similarly, ripening may be prolonged from 25 to 30 days in tropics to 60 days in temperate regions. Figure 2.4 shows schematic depiction of growth phases of paddy of a 120-day rice variety in the tropics, with onset of marked events in the life cycle of the plant.

The vegetative phase starts from the germination of seed, initiated by adequate absorption of moisture under favourable conditions of temperature (10–40°C). It starts with the emergence of radicle and plumule (coleoptile) through the husk. The first leaf breaks through the coleoptile on the second or third day. The leaves continue to develop every 3–4 days in early stage of this vegetative phase. Secondary adventitious roots soon develop and replace the temporary radicle and seminal roots, and form the permanent roots. At about 18–20 days, the seedling is ready to be transplanted to the field if the germination has been carried out in a nursery bed, a practice that is prevalent in Asian countries. Tillers soon emerge from the auxiliary buds at the basal internode and displace the leaf as they grow and develop. The primary tillers then give rise to secondary tillers. This occurs at about 30 days after transplanting. Tertiary tillers also emerge from secondary tillers as the plant grows longer and larger. The number of nodes on a tiller may vary from 13 to 16, with only upper 4 or 5 separated by long internodes. Stem elongation takes place in the later part of tillering stage.

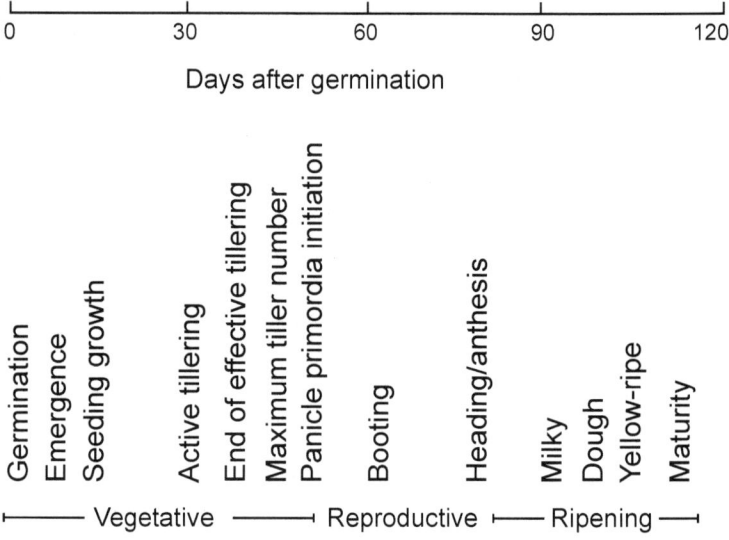

Figure 2.4 Schematic depiction of growth phases of paddy of a 120-day rice variety in tropics, with the onset of events during the life span of the plant [*Source:* Mclean et al, 2002, Adapted and used with permission from IRRI]

The reproductive phase begins with initiation of panicle at the tip of the growing shoot, and is visible to the naked eye after 10 days of the initiation. As the panicle continues to develop, the spikelets become distinguishable. The increase of panicle in size and its upward extension inside the flag-leaf sheath causes the leaf sheath to bulge. This bulging of the flag-leaf sheath is called booting. At this stage, senescence (aging and dying) of leaves and of non-bearing tillers are noticeable at the base of the plant.

Spikelet anthesis (or flowering) begins with full emergence of panicle, on the same day or on the following day. It takes 10–14 days for a rice crop to complete heading as there is variation in panicle emergence among tillers of the same plant and among plants in the same field. Heading stage is usually defined as the time when 50% of the panicles have emerged. Anthesis normally occurs from 10 am to 1 pm in tropical environments and fertilization, with self-pollination, is completed within 6 hours. Very few spikelets have anthesis in the afternoon, usually when the temperature is low.

The ripening phase follows fertilization and can be subdivided into milky, dough, yellow-ripe and maturity stages. These terms are primarily based on the texture and colour of the growing grains. The length of ripening varies among varieties from about 15–40 days, depending on the variety and climatic conditions, being shorter in the tropics and longer in cool temperate regions. As the grain passes through these stages, the senescence of the leaves on the tillers continues upwards. At maturity, the grains are fully developed, are hard and yellow. The field starts to look yellowish. Towards the end, the upper leaves also dry rapidly and considerable amount of dry leaves accumulate at the base of the plant. The crop is ready to be harvested.

2.4   Rice culture

2.4.1   Cultivation systems

A hot and humid climate, with a temperature range of 21–35°C, with clay and clay-loam soil is most suited for rice. Although tropical zone, and also the contagious subtropical regions provide the best climate, rice is grown under very diverse soil and climatic conditions across the globe. Unlike most other cereal crops, rice benefits from standing water. It is benefited most from the high rainfall climate and irrigated low-land cultivation system. About half of the global rice area is covered by such environment. It produces, however, more than 75% of the world's rice supply. Irrigated rice, therefore, remains the keystone to global rice. It is grown mostly with supplementary irrigation in the wet season and is reliant entirely on irrigation in the dry season.

Although majority of the rice varieties are grown in low land with full or partial irrigation, varieties have been evolved to suit rain-fed areas, called dry-land or up-land varieties and also for the deep water conditions. About 19% of the world crop is grown under rain-fed (non-irrigated) low land areas, and another 4% is raised in the rain-fed upland ecosystems. Because rice is grown under such diverse conditions, one set of cultural practices cannot be used effectively for all conditions. The practices followed for raising the crop under each of the above systems, therefore are different.

In the wet-land irrigated system, rice is grown in different ways. In the most prevalent practice, seeds are first sown in a nursery and the seedlings then transplanted to the field. This system is followed in the south and Southeast Asian countries. Seeds are sometimes directly put in the puddled soil, or drill-seeded into the dry soil, or seeded directly into water.

In the transplanting system, the seedlings are transferred from nursery to field in puddled soil. Water is allowed to remain in the field almost throughout the life of the crop, except towards the end. Puddling of the soil has many advantages. It greatly reduces the weed population and substantially increases the amount of water retained by the soil, thus reducing the losses through percolation. The requirements of manure vary depending on the varieties and soil conditions. The dosage of nitrogen for traditional indica type is low (40–60 kg per hectare), as they generally lodge (fall down) under heavy doses of nitrogen fertiliser. For japonica or the dwarf indica varieties, the maximum dosage could be up to 200 kg per hectare. The requirements of other two major nutrients, namely, phosphorus and potash vary from 50 to 90 kg per hectare depending on soil conditions, season and cultivation system.

In the dry-land system of cultivation, the seed is directly sown by broadcast (spread by throwing), with the outbreak of monsoon. The varieties suitable for this system are of short duration, usually of 90–110 days, and the yields obtained are generally low. The success of the crop entirely depends on sufficient rain fall. Total success is rare. Partial success is common. Sometimes the crop may totally fail, if the rains fail. Rain-fed rice environments, therefore, experience multiple stresses due to uncertainty in the timing, duration, and intensity of rainfall. Some of rain-fed rice areas are frequently affected by drought. The largest, most frequently, and most severely affected areas are the eastern India and north-eastern Thailand and Lao PDR. Drought is also widespread in Central and West Africa.

In the deep water cultivation system, the seeds are sown by broadcast on dry land with the early showers of the season. In some areas when the rainfall is more than 200 mm within that month, pre-germinated seeds are broadcast onto puddled fields without much standing water. The plants establish firmly before the flood water rises. The varieties suitable for this system have the capacity to lengthen rapidly and keep pace with the rise of flood water when the fields are flooded. So the panicles containing the grains remain floating above the water. The crop is harvested from boats. The deepwater rice is grown in tropical monsoon climate normally around river deltas and their flood plains, mainly in the backswamps and natural levees. This system is followed in the low-lying tracts of river Brahmaputra and Ganges in India and Bangladesh, in the Irrawaddy delta in Myanmar, in the Chao Phraya, and Mekong in Thailand, as also in Vietnam and Cambodia and parts of the Philippines. Some deepwater rice is also cultivated in West Africa in the Niger River basin, and in Ecuador.

2.4.2   Crop management

In the irrigated system of cultivation, which is most prevalent method of cultivation of rice, a nursery is raised first. About 50–70 kg of healthy and well-filled seeds, after treating with suitable fungicides, are sown in about 0.1 hectare of land. This is sufficient for planting over 1 hectare of main field. The nursery beds are thoroughly pulverised before sowing. Manure and fertilisers are applied as needed. Mixture of fungicides and insecticides are sprayed according to schedule. Beds are kept moist throughout the growing period and are properly weeded. Seedlings are lifted carefully without causing damage to the roots. Transplantation is generally done by manual labour. If the transplanting is to be done by machine, such as in Japan, the seedlings are grown in boxes. This requires a high standard of field and seedling conditions. Proper spacing between the seedlings while transplanting is also important. This depends on variety, type of soil, and time of planting. In general, a spacing of 15 cm between rows, and 10–15 cm between plants is desirable for early varieties. Spacing of 20×10 cm is suggested for medium and late varieties. Also, varieties with improved plant type and high tillering capacity are planted at wide range of spacing. Under poor or no weed control, closely spaced rice competes better with weed. The seedlings are planted preferably in straight rows as it makes it easier for weeding, spraying insecticides and top-dressing with fertilisers. Planting should be done to a moderate depth so that only the roots are pushed into the soil. Deep planting causes late tillering leading to uneven maturity of grains.

Crop management on the field requires appropriate management of land, water, manure and fertilisers, and control of weeds, pests and diseases at appropriate time intervals. A proper preparation of the land to start with is a must by repeated ploughing of the field and puddling of the soil. A continuous irrigation of the field should be ensured by flooding with 5–7 cm of standing water. In rain-fed areas, the paddy fields often become dry and the crop suffers from vigorous moisture stress. Although rice can be grown under upland, low land, and deep water conditions, stable high yields occur only under continuous, optimum flooded conditions.

As regard to the application of manure and fertilisers, organic manure, at the rate of 5–10 tonnes per hectare, is applied first to the field, about a month before transplantation. In case of green manure, it is incorporated into the soil 3 weeks before transplantation. The application of fertilisers and their schedule has already been discussed above.

Control of weeds in the irrigated transplanted rice system, with soil puddling and flooding of the field, is easier than in the rain-fed transplanted rice. Hand weeding is commonly used in almost all areas in south and

Southeast Asia. Rotary weeding is practiced in some Southeast Asian countries although not in others. Herbicides are used most commonly in areas of inadequate labour and high wages, particularly in eastern Asia – Taiwan, Korea and Japan.

Control of insects and diseases of rice plant is an important aspect of crop management in the field. A large number of diseases and insects attack paddy crop at various stages. Insects do far more damage to paddy in the tropics than in the temperate region. Stem borer, brown plant hopper, green leaf hopper and gall fly are some of the serious insects in the paddy field. Control of insect pests in the tropics currently depends largely on the use of pesticides, although many traditional varieties have some resistance to one or more insect pests or diseases. Among the microbial diseases, blast, blight and smuts take heavy toll. For effective control of these diseases and insects, two to three sprayings of fungicides and insecticides have to be applied during the crop period. However, pesticides are not only costly to farmers, they can disrupt natural biological control, and are also damaging to human health and the environment. Efforts have been made therefore to integrate pest control methods by combining cultivar resistance with pesticides so that the load of pesticides could be reduced. Promising lines are being developed that combine broad genetic resistance to all major insects and diseases in Asia. The concept of 'Integrated Pest Management' is gradually becoming an important guideline for pest control in paddy fields.

2.4.3   Harvesting and threshing of paddy

In the tropics, paddy grains reach physiological maturity about 30 days after heading or flowering; the grain becomes hard and relatively dry. At this stage the entire plant turns yellow and starts drying. Watering of the field is also stopped and the crop is ready to harvest. Many different systems of harvest have been devised for paddy depending on the environmental, cultural, religious and economic factors. A major portion of the paddy in the Asian countries, and that forms the majority of the paddy produced across the world, is harvested by hand sickle, though various types of knives are also used. Harvesting by combines includes both harvesting and threshing (separation of grains from panicles of the plant). This is mainly practiced in industrialised countries where farm holdings are large and labour is expensive. For manual harvesting, the straw is usually cut with sickles about 15–25 cm above the ground. In some areas of Indonesia and Philippines, the panicles are clipped off with a sharp knife. After cutting, the stalks are laid in small sheaves and bundles on the ground, stacked or hung on racks, to dry the ears for 2 or 3 days, prior to threshing or storage.

In the traditional practice, when paddy is harvested (the straw and the grain turn almost yellow) the average grain moisture is around 17% or less. It is found that by leaving the crop up to this dead ripe stage, there may be a small gain in the paddy yield. However, studies have shown that losses due to shedding, and also due to an increase in breakage of grains during milling, nullify the above gain. There is an optimum level up to which the paddy should be allowed to stand in the field, so that it gets dried to an appropriate moisture level, to achieve an optimum yield of milled rice with minimum breakage of grains when the paddy is milled. Losses ranging from 5% to 15% are estimated to take place due to wrong and late harvesting and threshing practices. On the other hand, harvesting of crop too early also leads to losses in yield and breakage due to soft immature grains. Paddy harvested at an average grain moisture content of 20–24% has been found to give maximum yield of good quality grains. At this moisture range, the paddy usually contains about 2.5% to 7% greenish grains and 0.4% to 1% milky grains. Counting of such grains in a representative sample, therefore, could serve as a simple, on-field, test for deciding roughly the proper harvesting stage of the crop.

Harvested crop has to be immediately threshed and dried under controlled conditions for producing grains of good milling characteristics. Threshing involves separating the grain from panicle. Threshing of hand-harvested sheaves requires adequate pre- or post-threshing drying. Manual threshing is done by beating the paddy heads on perforated platform made of bamboo. Mesh screen or other hard surfaces are also used. In some areas, plant shoots with panicles are held on the surface of small rotating drums with spikes. Occasionally, threshing is done by having animals tread on the harvested rice crop, or engine-powered small threshing devices. Harvesting by combines is practiced for large farm holdings, especially in the industrialised countries. Modern combine harvesters, such as those used in America, Europe and Australia can cut the straw and thresh the grain which is then separated from the straw, cleaned, and stored temporarily in self-contained bins. A high-capacity combine can harvest a crop over 6–12 hectares per day. Artificial dryers of matching capacities are essential in such cases to ensure good milling quality of rice.

In all cases, however, preliminary cleaning of foreign material from the grain and dying to safe moisture level, to about 13–14% is accomplished before paddy is placed in storage structures.

2.5   Rice production

With a total of 694 million hectares under cultivation, cereals represent more than half of the harvested area of the world (in 2010). They are the most important food source for human consumption and are the major supplier of energy and nutrients for the human population. Of the approximately 2.3 billion tonnes of cereals produced in 2010, roughly 1 billion tonnes were destined for food use, 750 million tonnes were employed as animal feed, and the remaining 500 million tonnes were processed for industrial use, used as seed or wasted. Unlike other cereals, rice is almost totally used for food purposes. Although 90% of rice is grown and consumed in rice-producing countries in Asia, it is also emerging as an important staple in some West African countries in recent years.

Table 2.1 World production of cereals during 2012–13 (estimate) [*Source:* USDA data]

Cereal	Quantity produced (million tonnes)	Proportion (%)
Maize	989.2	36.5
Paddy	713.2	26.2
Wheat	714.7	26.3
Barley	145.3	5.4
Sorghum	60.9	2.2
Oats	23.6	0.9
Rye	15.8	0.6
Other cereals	45.5	1.7
Total	2708.9	100.0

Table 2.1 presents the global production of cereals in 2013. Of the total production of a little over 2,700 million tonnes, maize, with about 1000 million tonnes, is the largest grain produced, representing a share of nearly 37%. Paddy and wheat, produced in equal quantities of a little over 700 million tonnes each, have a share of 26% each. Other minor cereals, with a total production of a little less than 300 million tonnes, contribute to the remaining 11%. The three major crops, viz. wheat, rice and maize, have shown continuous increase in the production over decades, depicting efforts to keep pace with the demand of growing world population. Of these three major cereals, maize has shown the highest growth of 44% in production over the last decade. For the same period, paddy production grew by 25%, while wheat by 16%. As the major part of maize use is in feed, starch, and related industries, the trend shows a higher rate of industrialisation, which is evident even in developing countries in the present times.

Table 2.2 World production of paddy [*Source:* USDA data]

Year	Area harvested (Million hectares)	Paddy produced (Million tonnes)	Yield (Tonnes/ hectare)
1984–85	144.1	464.9	3.23
1989–90	147.8	510.3	3.45
1999–00	155.9	608.6	3.90
2004–05	151.8	596.6	3.93
2005–06	153.9	621.4	4.04
2006–07	154.5	624.7	4.01
2007–08	154.9	642.0	4.14
2008–09	158.2	667.7	4.22
2009–10	155.9	656.2	4.21
2010–11	157.8	670.4	4.25
2011–12	159.2	694.5	4.36
2012–13	158.2	701.1	4.43
2013–14*	161.4	717.3	4.44

*Estimate

The global production of paddy over the 30-year period, from 1984–85 to 2013–14, is shown in Table 2.2, including data on the area covered and the yields obtained in the respective years. A steady growth in the area harvested and the quantity of paddy produced from it could be seen. It may be mentioned here that whereas the area under rice (paddy) increased by 12% over this 30-year period, rice production increased by 54%. During the same period, the world population also rose, by about 45%, from 4.83 billion to 7 billion. In Asia, which would be a better comparison model, the population rise was higher, viz. 52%. The production, therefore, has been keeping pace with the population growth so far. This has been possible with the adoption of high-yielding varieties, and use of better crop management practices. The current rice area is at an all-time high at around 160 million hectares compared with 120 million in the early 1960s. Further expansion of rice area in the future is not a viable option for most Asian rice-growing countries, where additional land is no longer available, and pressure on the existing rice land from urbanization and other non-agricultural uses is growing rapidly. For example, China's rice area has declined by more than 5 million ha (15%) in the past three decades and this downtrend may continue. Without the possibility of further area expansion in the future, yield growth will have to be maintained at 1.2–1.5% to be able to meet growing global needs. Yield improvements in the future need to be achieved in the face of the emerging constraints such as land and water scarcity, including depletion of ground water, environmental degradation, and rising input prices.

Table 2.3 Major (top 15) rice-producing countries of the world [*Source:* USDA data]

Country	Milled rice produced (million tonnes), during						Yield of paddy (tonnes/ha)
	2008–09	2009–10	2010–11	2011–12	2012–13	2013–14*	2013–14
China	134.3	136.6	137.0	140.7	143.0	144.0	6.72
India	99.2	89.1	96.0	105.3	104.0	106.0	3.64
Indonesia	38.3	36.4	35.5	36.5	37.5	37.7	4.72
Bangladesh	31.2	31.0	31.7	33.7	34.0	34.5	4.38
Vietnam	24.4	25.0	26.4	27.2	27.6	27.8	5.76
Thailand	19.8	20.3	20.3	20.5	20.2	20.5	2.84
Myanmar	11.2	11.6	10.5	10.8	10.7	12.5	2.65
Philippines	10.8	9.8	10.5	10.7	11.4	12.2	3.90
Brazil	8.6	7.9	9.3	7.9	8.2	8.5	5.09
Japan	8.0	7.7	7.7	7.6	7.8	7.7	6.73
USA	6.5	7.1	7.6	5.9	6.3	6.8	8.62
Pakistan	6.9	6.8	5.0	6.2	6.0	6.7	3.59
Egypt	4.7	4.6	3.1	4.2	4.7	4.9	8.95
Cambodia	4.0	4.1	4.2	4.3	4.6	4.9	2.49
Korea, South	4.8	4.9	4.3	4.2	4.0	4.2	6.76
Others	36.0	37.7	40.2	40.1	40.2	41.7	
World total	448.7	440.6	449.3	465.8	470.2	480.6	4.44

*Estimate

Rice is produced in 115 countries across the globe, and as already mentioned, under varied agro-climatic and geographical areas. However, only 32 countries produce more than 1 million tonnes each, of which 20 produce more than 2 million tonnes, and 12 of them produce more than 10 million tonnes of paddy each, annually. Presently, the top ten rice countries account for 85% world's production, their production values ranging from about 8 million tonnes to 144 million tonnes. Among these, 9 are from Asia, while one, Brazil, is from Latin America. Table 2.3 shows the country-wise production performance of the top 15 rice-producing countries of the world for the recent 6-year period, from 2008–09 to 2013–14, and also the yield data for the year 2013–14. It could be seen that China is the largest producer, contributing to 30% of world rice output, followed next by India with a share of 22%. Thus, these two countries together produce more than half of the world's rice. China, in fact, has 30% less area under rice cultivation than India but it produces 36% more rice. This has been possible on account of higher productivity in China, which has been attributed to the development and adoption of high-yielding varieties, (including hybrid varieties, which cover more than half of the rice area), improved crop management practices, better irrigation (almost all the area under rice is irrigated), and government policies with incentives to the farmers. On the other hand, India's major portion (55%) of rice area is rain-fed and hybrid varieties cover less than 4% rice area.

The next four countries in the top slot are also Asian countries, viz. Indonesia, Bangladesh, Vietnam and Myanmar (Burma), which produce between 20 and 40 million tonnes of paddy each, share together another 25%, taking the contribution of these six Asian countries to more than 77% of the world's produce.

There are eight countries which have recorded paddy yields of more than 6.0 tonnes per hectare. They are: China (6.72), Japan (6.73) and South Korea (6.76) from Asia, Italy (6.36) and Spain (7.53) from Europe, Peru (7.72) from South America, Egypt (8.95) from Africa, and USA (8.62) from North America. However, except China and Japan from these, others do not have sufficient area under rice cultivation, and their contribution to the world's produce is less than 1% each.

At the other end of the spectrum, at least seven countries produce less than one thousand tons per year, the lowest being Jamaica, which produces only around 2 tonnes of rice annually.

2.6   International trade of rice

2.6.1   Rice performance versus wheat and maize

Rice is a lifeline for many poor and small farmers. It is also a major food staple for large segments of the population. Governments in many developing countries have, therefore, intervened actively to stabilise domestic prices and promote self-sufficiency. For many developing countries, trade in rice largely remains as a residual option. It is not infrequent to see nations shifting from being a net importer to a net exporter, depending on the outcome of their paddy season. Concerns over scarcity of supplies have often led to the imposition of rice export limitation, including export bans, ceilings, taxes, etc. During 1990s, however, several countries liberalised imports and export policies, and allowed private sector to engage in rice trade, which has helped to make rice markets more open to foreign competition. In 2001, even a closed economy like China acceded to imports of rice.

The volume of rice traded internationally has traditionally been small, both relative to production and compared with the other major cereals, viz. wheat and maize. The volume of international trade for rice has increased from 7.5 million tonnes (4.4% of production) in the 1960s to 41 million tonnes in 2014 (8.2% of production). In contrast, the volume of trade in wheat and maize is much large. The 2014 figures were about 144 million tonnes (22% of its production) for wheat, and 130 million tonnes (13% of its production) for maize. The total volume of major cereals in the international market presently, for the year 2013–14, is nearly 330 million tonnes. Nearly half of it shared by wheat, while maize occupies about 40%, and rice contributes to the rest – a little over 10%.

As most of the rice tends to be eaten where it is produced, big volumes do not enter international markets. Maize and wheat, on the other hand, are produced worldwide but international trade is proportionately larger as demand has expanded, especially in Asian countries, due to changes in diets and increased industrial and feed use. It is interesting, however, to note that the percentage of produce traded in the international market for rice, though small, has been increasing slowly but steadily over the decades. It has doubled from 4.4% in 1960 to 8.2% in 2014. On the other hand, it has been hovering around 18–20% for wheat throughout the same period. In the case of maize, there was a surge from 7.5% in 1960 to about 19% in 1980s and 90s, but declined thereafter reaching to about 12–13% around 2000. The same percentage has been maintained in the subsequent years till now.

2.6.2   Rice exports

International trade of rice is characterized by a relatively small number of exporting countries interacting with a large number of importing countries, and is getting more concentrated over time. In the 1960s, the top five exporters had 69% of the world market. In the first decade of the 2000s, this share rose to 81%. The expansion of global rice trade from 1980 onwards has been mainly due to increased exports from Thailand, Vietnam, United States, Pakistan and China. Thailand, in fact has been consistently the highest contributor to international rice market, having a share of 25–35% up to about 2003. Its exports increased six-fold from 1.4 million tonnes per year in the 1960s to 8.4 million tonnes per year in the early 2000s, and surged further to 10 million tonnes in 2004, with a market share of 38%. Table 2.4 shows rice exports of the top 10 major players contributing to more than 90% of the total market during the recent 7-year period. Thailand maintained its top position till 2011, although its exports fluctuated between 7.5 and 10 million tonnes. Subsequently, it gave way to India, which became the topmost country, crossing 10 million tonne mark in 2011–12 and has maintained the same level in subsequent years. However, Thailand's exports renounced in 2013–14 to cross 10 million tonnes again. These two countries, therefore, shared together half of the export market. Vietnam, the second-largest rice-exporting country till 2010–11, slipped to third position due to rise of exports from India. Vietnam had become a major exporter during 1990s, following marketing and trade reforms adopted by the country at that period. Likewise, India's exports, which used to occupy fifth or sixth position in the global market, had surged upwards during 1990s and 2000, as a result of changes of policies, in particular the relaxation of the ban on ordinary (non-basmati) rice exports in 1994, and the concession of export subsidies between 2001 and 2003.

Table 2.4 International rice trade: Major exports (million tonnes) [*Source:* USDA data]

Exporting countries	Trade year						
	2007–08	2008–09	2009–10	2010–11	2011–12	2012–13	2013–14*
Thailand	10.0	8.6	9.0	10.6	6.9	6.7	10.3
India	3.5	2.2	2.2	4.6	10.2	10.5	10.0
Vietnam	4.7	6.0	6.7	7.0	7.7	6.7	6.5
USA	3.3	3.0	3.9	3.2	3.3	3.3	3.1
Pakistan	2.8	3.2	4.0	3.4	3.4	4.1	3.9
Cambodia	0.5	0.8	0.8	0.9	0.9	1.1	1.0
Brazil	0.6	0.6	0.4	1.3	1.1	0.8	0.8
Myanmar	0.6	1.0	0.4	0.8	1.4	1,2	1.3
Uruguay	0.8	0.9	0.7	0.8	1.1	0.9	0.9
China	1.0	0.8	0.6	0.5	0.3	0.4	0.3
Others	1.8	2.3	2.9	3.1	3.6	3.7	2.7
World total	29.6	29.4	31.6	36.2	39.9	39.4	41.2

*Estimate

The only non-Asian player in the major rice exporter's club has been the United States, with its consistent contribution of three to four million tonnes in the export market, on par with the contribution from Pakistan. The other countries which have shown notable contribution to the export market in recent years are Brazil, Egypt and Uruguay, having a share of less than a million tonnes each. China, which used to be one of the major contributors till 2003, showed a decline subsequently; her exports now reaching to 0.3 million tonnes. It has become, on the other hand, a major importer in the recent years, discussed later.

The high concentration of exports by contribution from only a few countries raises the possibility of disruptions in the market, leading to higher world prices that adversely affect net consumers in importing countries, but improve the welfare of rice net sellers. Conversely, exceptional production or subsidies on production in exporting countries could depress world prices to the benefit of rice consumers but adversely affect rice producers in importing countries. Perhaps most important, though, the high concentration of exports increases the probability of a production shock, or a change in trade policy in one or more of these countries could have a major impact on world market flows and prices.

2.6.3   Rice imports

In contrast to rice exports, imports of rice are widely dispersed across various countries. Imports by the five leading countries in the first decade of the 2000s (Philippines, Nigeria, Iran, Indonesia, and the European Union) were only 27% of the world total; the share of the top 10 countries was only 44%. This shows how thin the market spread is. Further, because of market segmentation, some of the larger rice importers have had major impacts on world rice prices. Indonesia's rice imports accounted for 10% and 15% of world trade in the 1960s and 1970s, respectively, which had major impacts on world rice markets. Large purchases by Indonesia even during 1997 and 1998, in the wake of El Nino weather anomaly, and by the Philippines in 2007 and 2009, are examples in which an individual importer contributed greatly to world price destabilization. Indonesia's rice imports were only 23,000 tonnes in 1993, which jumped sharply to 6 million tonnes in 1998, on account of sudden shortfall in domestic production. With many of the major rice importers hovering around self-sufficiency positions, they are a potential source of disruption on the international market. During recent years, China has emerged as the largest importer. The top 15 import destinations and their volumes during the past recent 7 years, i.e. between 2007–08 and 2013–14, could be seen in Table 2.5.

Table 2.5 International rice trade – Major imports (million tonnes) [*Source:* USDA data]

Importing countries	Trade year						
	2007–08	2008–09	2009–10	2010–11	2011–12	2012–13	2013–14*
China	0.3	0.3	0.4	0.6	2.9	3.0	3.5
Nigeria	1.6	2.0	2.0	2.6	3.4	2.9	3.0
Indonesia	2.0	0.2	1.2	3.1	2.0	1.5	1.0
Iran	1.6	1.5	1.5	1.9	1.7	1.5	1.6
Iraq	1.0	1.1	1.2	1.0	1.5	1.4	1.4
Saudi Arabia	1.2	1.1	1.1	1,1	1.2	1.2	1.3
EU	1.5	1.4	1.2	1.5	1.3	1.2	1.4
Philippines	2.5	2.0	2.4	1.2	1.5	1.5	1.4
Ivory Coast	0.8	0.8	0.8	0.9	1.4	1.2	1.2
Malaysia	1.0	1.1	0.9	1.1	1.1	1.0	1.1
Senegal	0.9	0.7	0.7	0.8	1.2	1.0	1.1
South Africa	0.6	0.7	0.7	0.9	0.9	1.0	1.1
USA	0.6	0.7	0.6	0.6	0.6	0.7	0.7
Brazil	0.4	0.6	0.8	0.6	0.7	0.8	0.7
Japan	0.6	0.8	0.6	0.7	0.6	0.7	0.7
Others	13.2	14.7	15.5	17.7	17.1	18.0	20.0
World total	29.8	29.4	31.6	36.2	39.2	38.6	41.2

*Estimate

There has been, however, a stable demand from African countries of Nigeria, Senegal, Ivory Coast (Cote d'Ivoire) and South Africa, and also the Near East countries of Iran, Iraq and Saudi Arabia. In these regions, rice imports in the recent past have been satisfying more than 40% of domestic requirement. The contribution is still higher for Central America and Caribbean countries, with half of their rice consumption now consisting of imports. For the developed countries as a group, imports satisfy about one quarter of domestic utilisation in rice. Despite this, the rice market is considered as being highly fragmented. Unlike rice exports, the geographical concentration of rice imports remains rather weak and thin.

2.6.4   Sub-sectors in rice trade

One reason why one particular type of rice is not acceptable on global level for marketing is the fact that there is strong consumer preference for specific types and qualities of rice, which is entrenched culturally in the respective rice-consuming markets. This limits the scope of substitution, and has led to the fragmentation of the market. In turn it has thwarted and delayed the establishment of internationally recognised grades or standards. With the result that there are more than 50 different published international price quotations for rice presently.

The international rice market can be broken down into several sub-markets, depending at least on three criteria, viz. the variety, the milling quality, and the processing.

As for the *variety* (it is a misnomer though), there are four distinct classes; (a) Indica – long grain rice (some short but slender grain rice may also be included), cooking to a firm and fluffy-textured product.; (b) Japonica – medium grain rice, sticky and moist when cooked; (c) Aromatic (this in fact is sold by the actual name of the variety, like Basmati and Jasmine, etc.) – normally long grain, scented variety; and (d) Glutinous (or waxy) rice – fully opaque grains, very sticky and moist when cooked.

Each of the above can be further distinguished according to the quality, mainly milling. FAO has arbitrarily classified rice containing less than 20% broken grains as "higher quality" and rice containing 20% or more broken grains as "lower quality".

The degree of *processing* is another criterion for categorisation of rice being marketed. Rice could be traded either in the in the form of paddy, husked rice (brown rice), milled rice or parboiled rice.

FAO has recently analysed the structure of the international rice market to see the spread of rice going through the above sub-sectors classification during 1992–94 and 2001–03. The data reveal that 75–76% rice traded was indica. The japonica had a share of 12–14%, the aromatic 9–12%, and the glutinous 1%. For aromatic rice, mainly Basmati, the destinations were the European Union, USA, Canada and Saudi Arabia. Major destinations for Jasmine rice (Hom Mali) were USA and China. However, large deliveries of Jasmine, with high percentage of broken rice, were imported into Africa, in particular to Cote d'Ivoire, Ghana and Senegal.

Considering the quality (milling), 75–77% of the traded rice was of high quality, while 23–25% fell into the low quality group.

As for the degree of processing, milled rice was the largest form in which the rice was traded, which comprised of 77–82% of the total market volume. Parboiled rice formed the next chunk with 13–15% share. Small quantities of paddy, 2–4%, and husked rice (brown rice), 3–4%, were also traded.

A strong import demand was also noted for the lower quality rice and parboiled rice by western African countries of Senegal, Cote d'Ivoire, and Nigeria, as also from South East Asian countries of Indonesia and Philippines.

The data thus show that aromatic rice varieties, lower quality rice and paddy have made large in-roads into the rice market and have established their shares.

Further reading

- Chang and Li (1980)
- Glaszmann (1987)
- GRiSP (2013)
- Maclean, Dawe, Hardy, and Hettel (2002)
- Lu and Chang (1980)
- Vergara (1980)

Recent views on the origin and domestication of rice has been summarised by:

- Fuller (2011)

3.1   Introduction

3.1.1   The necessity of drying grain

The importance of cereal crops, including rice (i.e. paddy or rough rice in this context), lies in their storability. Cereals are 'dry' grains. As a result, unlike other foods, they can be stored for relatively long periods of time without appreciable spoilage or deterioration. This property enables cereals and other food grains and nuts to be grown once (or may be twice) a year but to be eaten throughout the year. Cereal grains are thus food par excellence to provide food security. In other words, the power to provide food security by being able to be safely stored for long periods is the USP of cereals crops.

Clearly, drying becomes an essential step after the harvesting of cereal crops preliminary to their storage. Wet or damp food cannot be stored; damp food spoils quickly. Dry food can be stored. They do not spoil easily. Water is an enemy of food in that sense. When the seed is just fertilised, meaning when the new embryo of the new plant is just born, the material inside the seed cover, i.e. the future seed, is close to being entirely water. As new material is synthesised and the seed starts to grow, the solid matter in it goes on increasing and the moisture content steadily decreases. Finally, when the paddy grain attains maturity, approximately 30–35 days after flowering, the field having been previously drained of water, the paddy seed has a moisture content of approximately 30–35% (wet basis, wb). Of course, the actual moisture content differs depending on the prevailing weather. For example, the moisture may increase if there is rain or from exposure to dew during the night. Similarly, the moisture can go down to fairly low levels (even 20%) if the crop is exposed to strong sun during the day and the weather is dry. However, on the whole, we can say that the mature paddy crop has about 30% moisture at the time of attaining maturity. Then the crop dries gradually over a period of time especially if the weather is favourable. Obviously, this crop whenever harvested has to be quickly dried to a safe storage level before it can be stored. The ideal moisture content of paddy for storage is 12% (for long storage) to 14% (for short storage). This is discussed later.

3.1.2   Expressing moisture content

Moisture is an inherent part of any biological material including food grains, such as paddy. This moisture plays an important role in the use, processing and storage of the grain (in fact any biological material). The moisture content goes on changing depending on the prevailing atmospheric conditions. This is especially true in drying where the amount of moisture undergoes a continuous change. Therefore, one should have a clear understanding of how the moisture content is expressed.

The content of moisture in any material can be expressed in two ways: on wet basis and on dry basis (db). The moisture content of a material on wet basis (m_{wb}) expresses the amount of moisture contained in the material as it is, i.e. on the basis of the material as a whole including its own moisture:

$$m_{wb} = \frac{\text{Weight of moisture}}{\text{Weight of the material}} \times 100\%$$

$$= \frac{\text{Weight of moisture}}{\text{Weight of solid matter} + \text{Weight of moisture}} \times 100\%$$

The moisture content of the material expressed on dry basis (m_{db}) on the other hand expresses the amount of moisture contained by the material on the basis of its solid content alone:

$$m_{db} = \frac{\text{Weight of moisture}}{\text{Weight of the material} - \text{Weight of moisture}} \times 100\%$$

Each of these can be converted into the other:

$$m_{db} = \frac{m_{wb}}{100 - m_{wb}} \times 100\%$$

$$m_{wb} = \frac{m_{db}}{100 + m_{db}} \times 100\%$$

Both these expressions are useful and have specific applications.

The expression m_{wb} is useful to understand how much moisture the material as such contains. Its drawback is that the expression does not enable an easy and quantitative understanding of the extent of any moisture change that may have occurred in the material during a process. For example, when paddy is undergoing drying, its moisture content changes all the time. Now if the moisture is determined at two points of time, then the amount of change that occurred between these two points cannot be easily understood if the two moisture values are expressed on wet basis. For example, if paddy at 20% moisture (wet basis) is dried to 15% moisture (wet basis), then the loss in weight cannot be calculated as 5%. This is because, when the moisture content changes, in the wet basis expression, not only the numerator but also the denominator changes. The numerator is the weight of water and the denominator is the weight of the material, i.e. solid + water. As the weight of water contained in the material has changed, so the value of the denominator (which includes not only the solid matter but also the amount of moisture) has also changed. The two values cannot therefore be directly compared. In the same way, an equal weight of moisture (say 1%) removed from the material undergoing drying at two different points of time during the process cannot be directly expressed by equal change in the moisture value.

The above difficulty does not exist when the moisture is expressed on dry basis. After all the amount of solid matter in the material remains unchanged no matter the extent to which it has been dried. Therefore, the denominator in the m_{db} expression always remains unchanged whatever be the drying. The two stages can therefore be directly compared and the amount of moisture change can be directly calculated.

For this reason, m_{wb} expression is generally used in normal discussion. But for expressing any change such as during drying and for engineering purposes in general, the expression m_{db} is normally used.

3.2   Principles of grain drying

3.2.1   Hygroscopic properties of grain

Why and how can wet grain be dried and why and how wet or damp grain spoils are really related to some fundamental properties of grain (or food or biological material in general) in relation to its moisture. These properties are the hygroscopicity and equilibrium moisture content (EMC) and equilibrium relative humidity (ERH).

Hygroscopicity is an inherent property of all biological materials. Cereal grains, like all other foods and biological materials, are hygroscopic in nature, i.e. they exchange moisture with their surroundings. Grains lose moisture to the air if they are wet and the air is dry. They gain moisture if the grain is dry and the air is relatively wet or humid.

This is based on the property of vapour pressure of water. The moisture present in any material, including grain, generates a pressure of water vapour corresponding to the moisture content of the material. Similarly, the air has a specific vapour pressure of water depending on its moisture content, i.e. its relative humidity (RH). In this situation, water vapour flows from the grain to the surrounding air or from the air to the grain depending on whichever vapour pressure is higher. If the grain is wet and its vapour pressure is high, then water vapour flows from the grain to the air and the grain dries. On the other hand, if the grain is dry, i.e. its vapour pressure is low and the weather is wet, i.e. the air humidity is high, then the water vapour flows from air to the grain and the moisture content of the grain increases.

The grain and the surrounding air thus always exchange moisture depending on which one is at a higher vapour pressure and which one has less. Given sufficient time, the two eventually come into equilibrium. In this way, in a mass of grain, such as in a storage bin, the intergranular air and the grain mass eventually come into an equilibrium and the air within the mass attains a final humidity (ERH) corresponding to the moisture content of the grain. On the other hand, an amount of grain left exposed to the open atmosphere attains a moisture content (EMC) corresponding to the humidity of the surrounding air. This relationship is called the equilibrium relative humidity (ERH) (of the intergranular air) or the equilibrium moisture content (EMC) (of the grain). A general ERH–EMC curve is shown in Fig. 3.1. It will be noticed that the grain moisture increases rapidly at low relative humidity (RH) values, then relatively slowly at middle RH values, and rapidly again at high RH values. This ERH–EMC relationship is of fundamental importance in all aspects of grain drying, storage, spoilage and preservation. Figure 3.1 also shows that a relative humidity of 70–75% is in equilibrium with about 14% moisture in paddy, the highest limit for its safe storage.

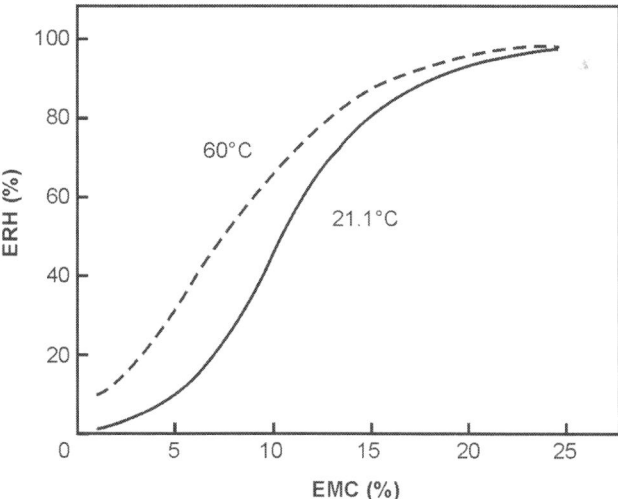

Figure 3.1 Equilibrium relative humidity-equilibrium moisture content relationship of paddy at 21.1°C and 60°C [Curves prepared from data compiled by Kunze and Caldarwood et al. (2004)].

3.2.2   The relation of hygroscopicity of grain to its drying

The relation of this hygroscopic property to drying is easy to visualise. It is precisely because of this property that clothes can be dried after washing. Clothes get saturated with water when washed. As they are then put on the line for drying in the sun, moisture flows from the clothes to the air because of the difference in the vapour pressure. So the clothes get dried. This is precisely what happens too when the grains are dried in the sun in the air. The grain is initially wet when it is spread on the drying floor or hung in stalk from a pole. Obviously, it has high water vapour pressure. The surrounding air with which the wet grain is in contact has a much lower vapour pressure. It is more so if the sun is shining. This difference in vapour pressure forces the water vapour to flow from the grain to the air especially if the grain mass is regularly raked and turned.

The same principle operates in spoilage of damp grain (or food in general) in storage too. The major part of a grain is food material stored by Nature to provide for the future growth of the seed during germination. But human ingenuity has so developed the agricultural system by appropriate selection and breeding that the overwhelming bulk of the grain crop is used as food for themselves rather than for the future embryo. Only a small amount of grain is kept and used as seed for agriculture. The grain meant originally to provide for the future embryo thus gets turned overwhelmingly into human food. However, the food for humans is also equally liked by other forms of life, such as microorganisms (especially fungi or mould) and insects and other predators.

So, when we want to store food for our future use, it must be done under such conditions that do not favour, in fact prevent, the growth of microorganism. It is this microbial growth that causes spoilage and destruction of the food. The property of hygroscopicity and the equilibrium relative humidity (EMC and ERH) becomes relevant here again. Microorganisms have got very specific relative humidity (and temperature) requirement for survival and growth. They cannot survive, or at any rate remain dormant, when the relative humidity is too low. They start to grow and flourish with exponential speed as the relative humidity (and temperature up to a limit) of the inter-granular air increases. Obviously, before one stores grain, it has to be dried to a moisture level where the EMC or ERH does not favour growth of any microorganism. Fungi or molds are always present in grain in a dormant state. In general, they cannot grow in an atmosphere of approximately less than 70% RH, i.e. in paddy having less than about 13% moisture. Mould grows in exponential rate beyond this RH (or moisture) if the temperature is not too unfavourable (i.e. unless the temperature is too low, say < 15°C).

3.2.3   The role of air in drying

The relation of the principles of ERH or EMC to drying of grain is now easy to understand. Had the grain material not been hydroscopic in nature, i.e. had it not been prone to exchange moisture with the surroundings, there would have been no scope for drying of the material at all. It is precisely because the grain, like all other biological material, is always prone to exchange its moisture with the surroundings that a moist or damp grain can be dried. The trick is therefore to manipulate its surrounding atmosphere as to make it natural for the grain to lose its moisture. In other words, if the grain is damp and the surrounding air has or can be arranged to have a low humidity, then circumstances are automatically created for the grain to dry.

Air plays a crucial role in this process. Four properties of air are relevant here. The first is the relative humidity of the air, i.e. its water vapour content. The lower the content of its water vapour or humidity, actually the relative content or the relative humidity (RH), the greater its ability or tendency to extract moisture from the damp grain.

The second air property is its temperature. The temperature plays a role in two different ways. On one hand, the relative humidity of the air for a given amount of absolute humidity, or water vapour content, depends on its temperature. The higher the temperature, the lower is the RH for a given amount of absolute humidity or vapour content and vice versa. What it means is that even when relatively wet, the ambient air can be made suitable for drying simply by heating or raising its temperature. Heating air to raise its temperature by 11°C leads to halving its RH. Thus if air with 90% RH and 25°C is heated to 36°C, then its RH is lowered to 45%. Heating thus automatically reduces the RH of the air and increases its ability to absorb moisture. This is precisely the reason we hang out washed laundry on a clothes line in the sun rather than in shade if the sun is shining. Air temperature also plays a second role. It helps in actual evaporation. Evaporation of moisture requires supply of energy (latent heat). Therefore, a higher air temperature facilitates drying by making energy for evaporation more easily available.

The third property of the air relevant to drying is its ability to take the water vapour away. It is not simply evaporation of moisture or the drying potential of the air that is enough for drying. After all, the evaporated moisture has to be removed. This is what is accomplished by the moving stream of air whether natural or artificial. That is why laundry dries faster when there is a good breeze rather than in a still atmosphere, other things being equal.

A fourth property of the air relevant to drying is its content of oxygen, although this role comes into the picture in some systems and not in others. In a drying system, where the air or the grain is being heated, the role of oxygen comes into the picture. It is the oxygen which enables a fuel to be burned to generate heat needed either to heat the air or indirectly the grain.

3.2.4   Psychrometry

The first two properties of the air in relation to its humidity and temperature mentioned above are called the psychrometric property of the air. These relations are well expressed in a chart called the psychrometric chart (Fig. 3.2). It is good to have a clear understanding of air psychrometry to have an insight into how and why a drying process operates. Even though a psychrometric chart may look rather complicated, its essence is really very simple. This is explained in the accompanying skeleton diagram (Fig. 3.3).

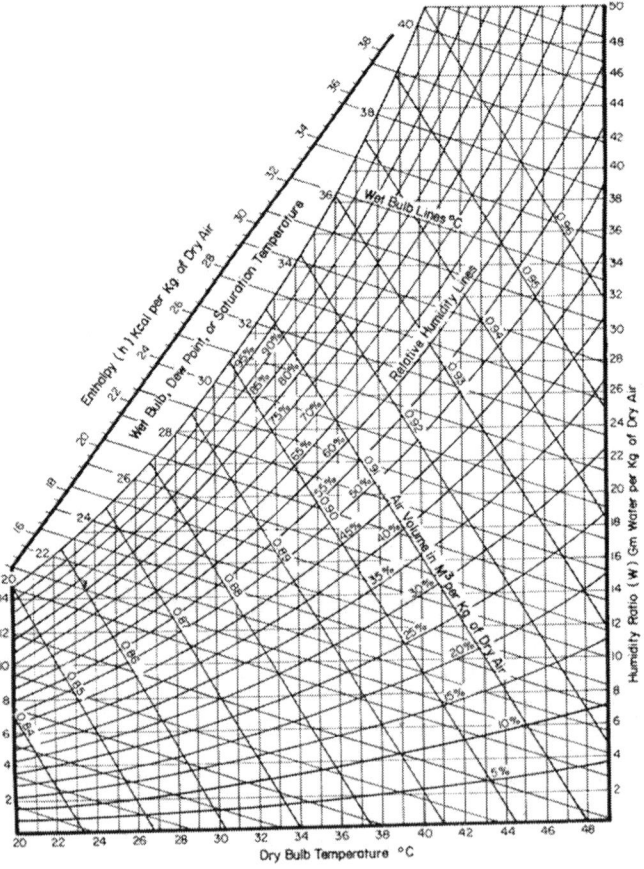

Figure 3.2 Psychrometric chart

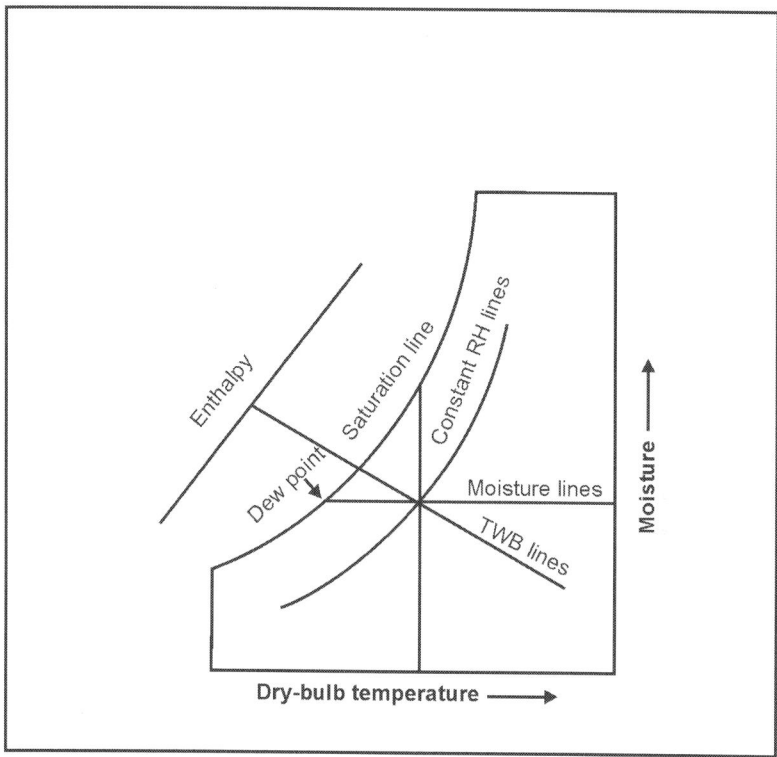

Figure 3.3 Skeleton psychrometric chart identifying important lines and scales

But before explaining this chart, a few indices of air properties should be first understood. First is the dry bulb temperature of the air (T_{DB}). This is the sensible temperature, i.e. the temperature as recorded by an exposed thermometer. The second is the wet bulb temperature (T_{WB}). This is the temperature as recorded by a thermometer when its mercury (or other fluid) bulb is covered by a thin film of water. When the bulb is covered by a thin film of water, the water would evaporate into the air depending on the capacity of the air to absorb water vapour. Clearly, more water vapour would be absorbed if the surrounding air is dry and less vapour or nothing would be absorbed if the air is highly humid. Now any evaporation of water into its vapour form involves the extraction of an appropriate amount of energy in the form of latent heat (540 kcal/g). This amount of heat has to come from the surrounding layer of air and the thermometer bulb which are thereby cooled to that extent. In other words, the bulb would now be at a lower temperature than the dry bulb temperature noted above. This is therefore called the wet bulb temperature (T_{WB}).

Therefore, this T_{WB} is always lower than dry bulb temperature (T_{DB}) except when the surrounding air is already saturated with water vapour. In the latter case, the air is unable to absorb any more water vapour, whereby it does not suffer any fall in its wet bulb temperature caused by moisture evaporation, in other words its T_{WB} is equal to T_{DB}. Clearly, there would be more fall in temperature when the air is dry and so has a high capacity to absorb more water vapour. The fall would be less when the air is rather wet and its potential for absorbing water vapour is low. The difference between the T_{DB} and T_{WB} is therefore a direct measure of the relative humidity (RH) of the air. The precise value of RH can be read from a chart of T_{DB} against T_{WB} at the specific temperature.

Let us now understand the essence of a psychromatic chart (Fig. 3.3). The bottom horizontal line is the line of sensible heat or dry bulb temperature of the air, T_{DB}. The T_{DB} increases from left to right. The right vertical line represents the absolute humidity of the air, i.e. the gram water vapour per kilogram dry air. The humidity content increases along the vertical line. The angular left to right downward lines represent the wet bulb temperature T_{WB}. The curved left to right rising lines represent the constant RH lines, and the parallel horizontal lines represent the dew point values shown on the left saturation line. The dew point is the temperature at which the air becomes fully saturated for the corresponding absolute humidity values shown on the right vertical line. The enthalpy or the total heat content of the air, comprising of the sensible heat plus the latent heat for the humidity content, is shown in the left cross straight line. A complete picture about all the properties of the air (its dry or wet bulb temperature, moisture content, RH, enthalpy, etc.) can be easily read from this chart once any of its two properties (usually T_{DB} and T_{WB}) are known.

What happens during drying is easy to follow from the psychromatric chart. For example, a sample of air at point A in Fig. 3.4 has a specific T_{DB} as shown in the horizontal line below and a RH as read from its intersection with the corresponding RH line. It has its RH drastically lowered if it is suitably heated to the point B. The enthalpy of the air is thereby increased from the original value of h1 to h2. This hot air now enters the dryer and meets the damp paddy. So this air now dries the paddy as far as it can and perhaps exits the dryer at point C (close to saturation point E but actually perhaps not fully saturated). The encounter of the hot air with the damp paddy is an adiabetic process, i.e. no energy is lost or gained. So, the air reaches the point C (or E). That is, the air loses some sensible heat but gains moisture, and correspondingly the grain loses the same moisture. This example shows how one can get a picture of what happens during drying if the initial T_{DB} and T_{WB} and the final T_{DB} and T_{WB} values of the air are known.

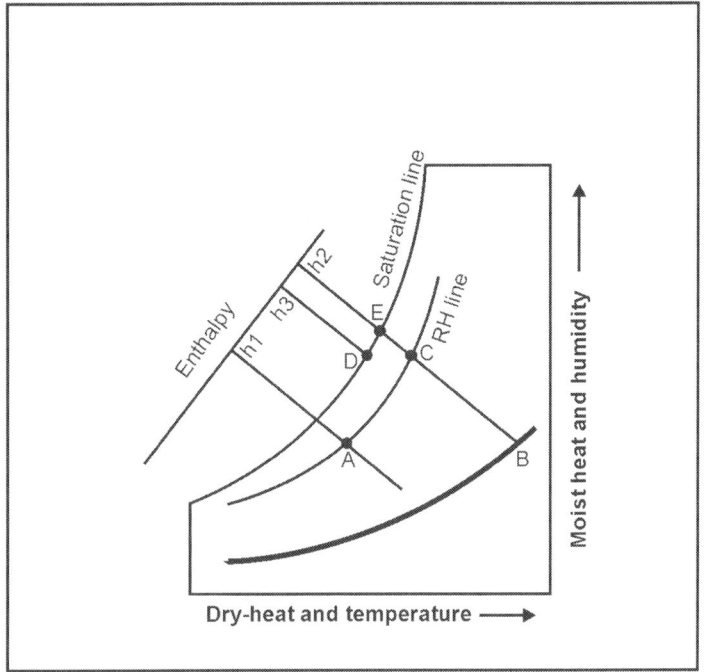

Figure 3.4 Psychrometric representation of grain drying

3.2.5   Theoretical model of drying

How exactly a process of drying proceeds and how the grain moisture content changes in quantitative terms with respect to time are important questions in drying. These aspects were studied by scientists initially in laboratory experiments, especially in the United States. These experiments were done primarily with maize (corn) but were also repeated with other grains later including paddy. Small amounts of grain were spread in a thin layer in an oven and then hot air was blown through the seeds. The amount of air, its temperature and the grain moisture content were the three variables whose effects were studied. The data obtained on the rate of fall of the moisture content thus determined showed that there were three stages of drying as shown in Fig. 3.5. Initially, there is a warm-up period (AB) when the grain and air come into an equilibrium. Then there is the constant rate period when the rate of moisture removal is constant (BC). This happens for a short period of time at the beginning when the grain is nearly saturated with moisture. This is followed by a period of falling rate of moisture removal (CD). The falling rate happens because evaporation of moisture can take place only from the grain surface, not its inner parts. Once the grain surface is no longer saturated with water, moisture from inner parts of the grain has to first migrate to the grain surface. This is called diffusion. Now the rate of diffusion is far lower than that of evaporation. Therefore, the rate of drying gradually falls as the drying proceeds and creates the period of falling rate. In actual drying practice of paddy, the period of constant rate is very small or marginal and the bulk of the drying occurs at the falling rate.

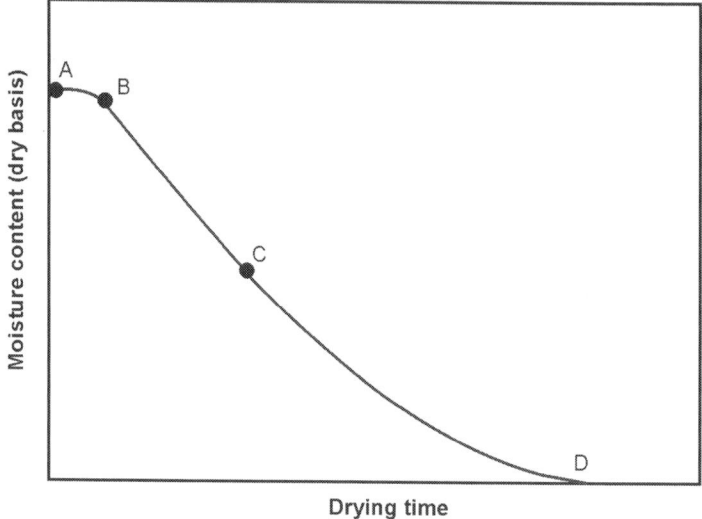

Figure 3.5 Rate of drying in thin-layer drying of paddy

The effect of the drying condition on the grain quality, especially on the development of cracks in paddy leading to its potential breakage during milling, was also studied during these experiments. This aspect is discussed later.

3.3   Drying systems

Potentially there are three systems of grain drying: (a) in ambient air, (b) by forced natural air, and (c) by forced heated air.

3.3.1   Basic approach to the three drying systems

Initially, grains were being dried with natural air where the grain was simply exposed to the ambient atmosphere. This practice is still being followed in most paddy-growing parts of the world. This is especially true of tropical countries where temperature conditions are favourable, and/or where subsistence or near-subsistence farming is still in existence. When the farm sizes are tiny, the amount of produce is small, capital and infrastructure are meagre, and the atmospheric conditions are favourable, the idea of using artificial or mechanical power for drying is obviously impractical. This factor is always not realised. Viewing the matter only from the point of theoretical efficiency without taking into consideration the scale of operation or the cost of capital is erroneous. The level of technology ultimately self-regulates and adjusts to the prevailing level of operation, infrastructure and capital.

In this system, the grain is partly left to dry in the stalk in the field itself to a fairly low level of moisture (15–18%). Or often, the just harvested paddy is left on the stubble for a few days to dry to a level of about 15–18% before being threshed. These practices have their obvious drawbacks in the form of grain cracking and potential breakage of rice during milling (to be discussed below). But the practice continues nevertheless because of the overall circumstances. Alternatively, if space is available, the partially moist paddy can be dried or at least finish-dried by spreading on the yard, preferably in the sun.

Whenever circumstances and available facilities permit and especially when the scale of operation requires, drying can be accomplished by using forced air. Here unheated or heated air is forced through the grain rather than exposing grain to natural air. A current of air generated by an appropriate fan is passed through a mass of stationary or moving grain. As a greater amount of air than the naturally circulating air is used in this system, drying is obviously accomplished in a lesser time or more grain can be dried in a given time. Obviously, this practice would be adopted not only when there is sufficient capital but also when the amount of grain to be handled justifies and requires the additional cost and facility and enables the cost to be absorbed. Here there are two possibilities. If the circumstances are appropriate, the air that is forced through the grain can be the natural (unheated) air. This is feasible if the amount of moisture to be removed is not very high, say when paddy having < 18% moisture is to be dried and the weather is not too humid.

As already discussed earlier, for safe storage of paddy, its moisture has to be reduced to less than 14% if the paddy has to be stored for less than say a few months, and less than 13% if storage is for longer period. This amount of moisture is in equilibrium with air of relative humidity of 70% or less. In other words, whenever the RH of air is more than 70%, that air cannot dry the grain to the desired level and may on the contrary in fact increase its moisture if circumstances are otherwise conducive. If on the other hand, the air is very dry (say the RH is less than 40%), and the paddy being dried is in a large bulk, perhaps in a bin or store, then drying would be improper. Such air would lead to overdrying, especially when the paddy is in large bulk, and may cause deterioration in quality (cracks, which is discussed later).

Drying a large bulk of paddy with forced natural air in a bin is thus somewhat tricky. First of all, the air has to have an appropriate RH. If its RH is too high, the air can be heated slightly to reduce its RH to an appropriate level. This is called supplemental heat, which should raise the air temperature by not more than 5–8°C. One may remember that heating of any air to raise its temperature by 11°C halves its RH. So, for the purpose of grain drying, the objective should be to heat the air only a little, so that its RH comes down to approximately 60%.

The above is drying of paddy by forcing natural or near-natural air through the grain. Here usually the grain is taken in a large mass in a bin which is used more for storage than as a receptacle to hold grain for drying. This is actually an intermediate stage of mechanisation. A still higher level of mechanisation can be, or may require to be, adopted, i.e. heated-air drying.

Obviously these different scales of operation and mechanisation are adopted not on the basis of a theoretical idea of efficiency but based on the required scale of operation and the cost of the factors of production (labour and capital). As a matter of fact, this high level of mechanised operation of drying of paddy, to be discussed below, was first adopted in the USA around or shortly after the World War II. Paddy was being harvested before that time either manually (in under-developed or semi-subsistence farming countries) or by simple reapers and binders. Combine harvesting, i.e. mechanised harvesting, threshing and transport, came into being around that time especially in USA and changed the scenario. Combine harvesting required that the plants were strong and erect, at which stage the grain moisture would have had to be rather high (25% or more). In addition, the grain was immediately threshed. Besides, the combined harvesting and threshing resulted in the job being done fast. The end result was that a very large mass of very high moisture threshed paddy became suddenly available for further handling. There was no alternative for this material than to be dried very fast to prevent its spoilage. This could only be done by drying with air heated to quite a high level. So heated-air drying was adopted out of necessity, not out of academic considerations of 'efficiency'.

3.3.2   Actual drying processes

Having explained the basic approach to different drying systems, the actual systems in use can now be described in brief.

Natural drying

The first is natural drying. Here the wet or damp paddy is exposed to natural air, especially in the sun wherever sunshine is available. Considering the prevailing atmospheric temperatures and sunshine, the system is feasible only in tropical areas. Apart from drying in the rack or directly on the stubble in the field after harvest, ideally this drying should be done in a drying yard after the grain is threshed. The main points to remember here are the following:

(i)   The grain should be spread in a relatively thin layer, at any rate not in a layer thicker than say 4 in., in order to avoid uneven drying.

(ii) The paddy should be regularly raked and turned so as to expose and bring the bottom layers to the top and expose the entire mass of the grain to the air and sun.

(iii) The grain should not be continuously dried for too long when the sun is shining very strong and the weather is hot. It should preferably be collected into a heap and kept aside for some time at such times. This action will work as a sort of tempering, the meaning and reason of which will be explained shortly.

(iv) Care should obviously be taken not to over-dry the paddy. In other words, the grain should not be dried more than down to about 13–14% moisture.

(v)   Another system can be adopted wherever feasible. In certain rice mills in India, where large-scale parboiling is being regularly carried out, the installed LSU dryers can be used for finish-drying of just-procured semi-damp paddy. In this system, the going parboiling process is temporarily suspended for some days or weeks. Then, the thus-freed LSU dryers are used to pass the semi-damp paddy through them while the fan blows unheated natural air through the paddy. Under appropriate conditions, this process enables 1–2 percentage points of moisture to be removed from the paddy.

In-storage drying

Natural drying as described above cannot be performed in countries in the temperate region where the weather is too cool. Wherever the quantity to be handled is not excessive, drying with forced natural air (or slightly heated air) can be performed in such conditions combined with storage of the grain especially by individual farmers. In this system, the storage bin or structure is so constructed as to provide a ducting system through which air can be forced by a fan through the grain. If the relative humidity of the air is appropriate (around 60–75% RH), then this action would obviously lead to drying of the grain in a measurable time (a few days to weeks). The low temperature protects the grain from deterioration during the period. The grain then continues to remain in the structure now serving as a storage bin.

A few issues are involved here. First, if the temperature is too low, then drying would be too slow and the process would take too long a time. Nonetheless, if the temperature is sufficiently low, once the mass is adequately cooled by the passage of the air, the grain even though remaining rather damp (say with 15–18% moisture), would not necessarily get spoiled too quickly at such temperatures. So the system can be operated nonetheless. Second, the fan must be stopped whenever the humidity is too high (over 75% RH). Alternatively, in such a situation, in prolonged bad weather, the air can be heated by a few degrees (supplemental heat mentioned above) to lower its RH appropriately. Third, in this system, since the air has to pass through a large mass of paddy, all the grain mass does not get dried simultaneously. The air starts to dry the grain at the point of entry, continues to travel through the mass as it soon gets saturated with moisture or near saturated, after which no further drying would occur in the remaining mass of paddy even though the air is passing. In other words, a drying zone slowly passes through the mass in this system, the previous zone being dried already and the following zone being still wet (Fig. 3.6). Fourth, the system should not be operated if and whenever the ambient air is too dry (or too much supplemental heat should not be used, whereby RH is lowered too much). In such a case, the grain would be dried down to its EMC value, which would be too low if the RH of the air is too low. In other words, the paddy should not be allowed to be dried down to uneconomical levels (below 12% moisture).

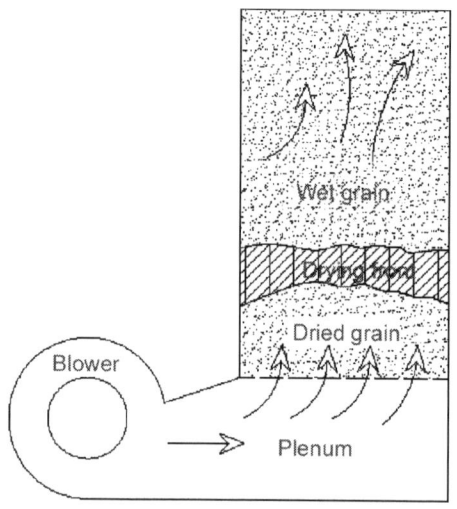

Figure 3.6 A drying front passes through the grain mass during in-storage drying of grain in a bin

These drying-cum-storage structures can be of various kinds and can come in various shapes (Fig. 3.7). It can be either a vertical, cylindrical bin, or a quonset, or elevated warehouse-shape structure with an appropriately slanting roof to match the angle of repose of the grain, or it can be a regular storage warehouse. Whatever be the shape, the main point is to provide an appropriate and adequate ducting system, so that the air forced by a fan passes uniformly through the entire mass of paddy in all corners. The air entry can be either at the bottom or through the top. Both of these have their advantage and disadvantage. Forcing air from the bottom has the advantage that the end of the process can be easily determined. When the grain at the top layer has just dried, one knows that the entire mass has been treated. On the other hand, its disadvantage is that there may be condensation followed by mould growth in the top layer if the atmospheric temperature is too low. Pulling air from the top and make it exit through the bottom does not have the disadvantage of possible condensation. At the same time, if the weather happens to be rather warm, it has the possible danger of drawing the warm layer of air at the top through the grain mass, and thus warming the grain.

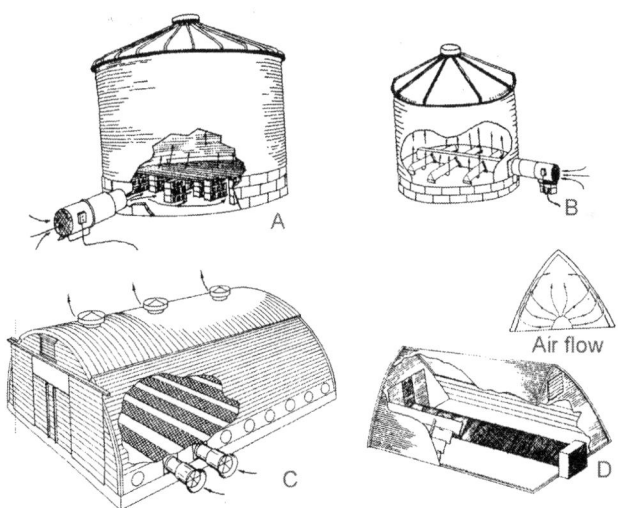

Figure 3.7 Different types of bin dryers [Reproduced from Dachtler (1959)]

Finally, it should be noted that the fan capacity is a crucial issue in such systems. The power requirement of the fan increases exponentially with the height of the grain mass and also with increase in the volume of air being passed through. The fan must develop adequate static pressure to enable the air to pass through the grain mass. This static pressure requirement increases with the depth of the grain mass. The cost of the fan and its power requirement increase drastically with these requirements and become the most important part of the cost of the entire system.

Heated-air drying in continuous-flow dryers

There are situations when the amount of wet paddy to be handled is very large. This happens especially under combine harvesting system wherein heavy amounts of high-moisture threshed paddy start to arrive in bulk in a central drying station. A system of fast drying has to be adopted in such cases. Natural drying or in-bin drying would not be able to handle such situations. Such paddy has to be handled by continuous-flow, heated-air drying.

In this system, the air is heated to high temperatures (> 50°C) whereby its RH is reduced to a very low level. This low RH combined with its high temperature converts the air into a potent dehydrating vehicle. A stationary bed of paddy cannot be dried with such air. In that case, the paddy would be reduced to its EMC value which might to be as low as perhaps 4–6 % moisture. In other words, in such systems, a moving bed of a relatively thin layer of wet paddy (5–8" thick) is exposed to hot air for a short time so as to reduce its moisture level by a small amount at a time.

Three such types of dryers are in operation. Two are columnar dryers wherein the damp paddy is lifted to the top of the structure and allowed to fall by gravity between two screens placed 5–8" apart (Fig. 3.8). Heated air is introduced through a plenum and passes through the screen, thereby drying the paddy. These columnar dryers can be either non-mixing, where the paddy flows from top to bottom in an undisturbed layer. Or it can be mixing, wherein some baffles are built in to the channel, whereby the paddy gets mixed together as it tumbles down. The second system is more efficient as it allows for even drying. The paddy is somewhat unevenly dried in the first system. This is because the inner layer is first to meet the hot air and so dries more than the exit layer, where by that time the air would have cooled a little and also picked up some moisture. The structure may be 40–50 feet high and some 15–20 feet wide.

The third dryer is called the LSU (Louisiana State University) dryer (Fig. 3.9). This is a box wherein several layers of inverted V channels are provided. Each alternate layer of Vs is open to the inlet air to allow the hot air to enter and closed at the opposite end. The next layer of Vs is open at the opposite exit end and closed at the air entry side to allow the used air to exit. Here also good mixing takes place because each alternate layer of V channel is placed slightly offset from its top and bottom layers. As a result, the down-coming stream of paddy gets divided and mixed together, while hot air rises and passes through the layer of grain into the exit channel. This LSU dryer is probably the most widely used by the industry.

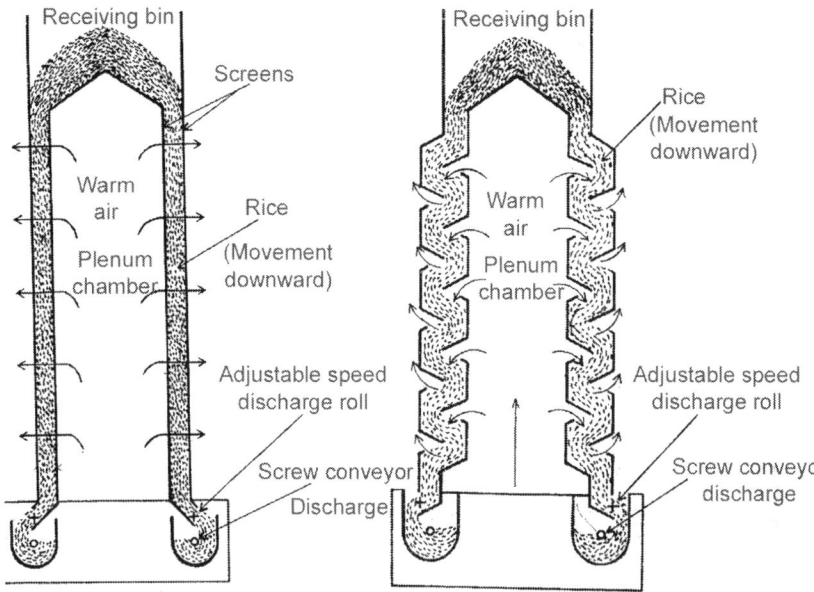

Figure 3.8 Non-mixing- and mixing-type columnar dryers [Reproduced from Dachtler (1959)]

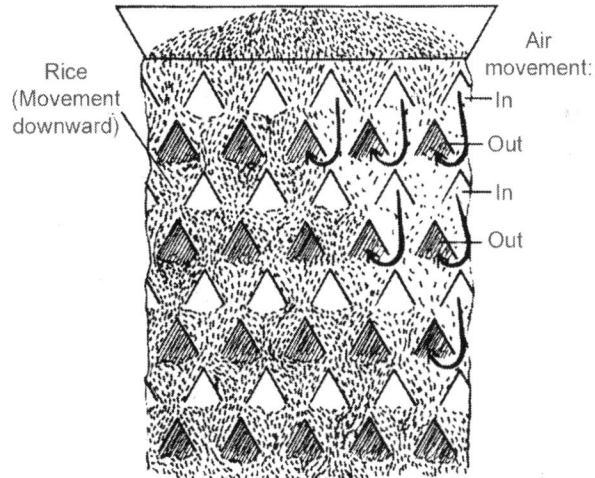

Figure 3.9 Louisiana State University-type continuous-flow, heated-air dryer
[Reproduced from Dachtler (1959)]

Two additional innovations in this system have been introduced since 1980s and 90s. The need for the innovations arose out of the need to increase the capacity of the drying systems. The objective was to meet the demands of increasing production and hence arrival of too much quantity of wet paddy to be handled. The first innovation consists of adding an aeration system into the tempering bins. In heated-air drying, drying is in any case performed in several passes, combined with intermediate tempering between the passes, as explained in detail below. It has been found that the paddy, slightly warm after each pass, can be made to undergo a little additional drying during the tempering if the grain is exposed to a low level of aeration in the tempering bin. As the rate of aeration is kept low, the additional drying is not high, thus avoiding the possibility of grain damage. At the same time, the small amount of drying (reduction of moisture by 0.5–1%) helps in the overall process without adding much cost.

The second innovation based on the same principle is to finish-dry the grain in a bin. That is, the paddy is dried by the regular system (i.e. in continuous-flow, heated-air, multi-pass system) down to about 16–18% moisture. It is then collected in a huge storage bin provided with efficient ducting systems for uniform passage of air. There the paddy is finish-dried to safe storage levels using low levels of natural-air aeration. This is based on the principle that heated-air drying is most effective (for drying) and least damaging to grain when grain moisture is high. It is the reverse when the moisture content is low: that is, drying becomes slow but damage high in hot-air drying towards the fag end of the process. It is best at this stage, therefore, to resort to slow drying with natural air.

3.4   Relation of drying to rice damage and how to prevent it

In drying of other grains or food material, the main concern is how to evaporate and remove its excess moisture. That is, it is mainly a question of setting up a system of a proper mix of aeration, humidity and temperature, such that the excess moisture in the material can be evaporated and taken away. The situation is more complicated when it comes to drying of paddy. This is because of two reasons. First, unlike other grains, rice is mostly consumed in the whole grain form after cooking. Therefore, maintaining the integrity of the grain throughout its handling and processing is of paramount importance. Any breaking of the grain during its milling to produce milled rice from paddy is highly undesirable. It is precisely here that the second difficulty arises. Drying under certain situations can and does cause damage to the rice grain, thereby making it liable to get broken during its subsequent milling. So in the case of drying of paddy, it is not enough to design an appropriate system of moisture reduction. The process must also be performed in such a way as to avoid damaging the grain.

It has been mentioned above that the lion's share of drying occurs at the falling rate stage. The falling rate happens because actual evaporation and removal of moisture occurs only from the surface of the grain undergoing drying. The moisture in the inner layers of the grain has to first travel to the grain surface by diffusion and only then can it evaporate. Now diffusion is a much slower process than evaporation, especially under rapid drying with heated air. The net result is that during heated-air drying, a moisture gradient develops in the grain undergoing drying and progressively increases in intensity as drying proceeds. The surface gets more dried, while the inner layers remain more moist. This differential drying in different layers of the grain sets up a stress within the grain. As the stress increases with continued drying, beyond a certain level, the grain releases the strain by cracking or fissuring. Such cracked or fissured grains then become liable to break when the paddy is subsequently milled to produce white rice.

The above phenomenon clearly means that the drying of paddy is not simply a question of removing or evaporation of moisture. It is also a question of carrying out the moisture removal in such a way as not to damage the grain. These aspects have been widely tested in laboratory experiments to find out the conditions under which the grain does crack or does not. It has been ascertained that there are precise conditions when alone the grain cracks. That is, if the drying is too fast or if the quantity of moisture removed in a unit time is too high, that is when the conditions become ripe for fissuring. Clearly, drying with natural air either in natural systems or in in-bin systems mentioned above are not such as would promote grain damage. Cracking of rice grain would occur only under heated-air drying or in drying in strong sun. The way to prevent fissuring is also evident from what has been discussed above. If the rate of drying is not fast enough, or if the amount of moisture removed in a unit time is not too high, then cracking would not occur.

Since fast drying is incumbent under conditions explained above, what it means is that the amount of moisture removed in unit time must be kept within strict limits. It means that heated-air fast drying has to be performed in several stages rather than in one go. This is called pass drying because each passage of paddy through the dryer is called as one pass. So, a small amount of moisture is removed in one pass, after which the partially dried paddy is allowed to rest, which is called tempering. During tempering, evaporation from grain surface stops but diffusion of inner moisture continues. Thereby, the moisture content in the grain eventually equalises after some time, at which time drying can be resumed again. The required time for tempering is short (4–8 hours) if the paddy is slightly hot or warm and longer if the paddy has been cooled (12–24 hours). So, the trick in such situation is to do the drying in several stages or passes with intermediate tempering between two passes. If properly done, damage to the grain can be totally avoided even when the actual in-dryer drying is done very fast. This way an installation with even one or a few dryers along with several tempering bins can be able to handle a large mass of damp paddy safely without damage or microbial deterioration.

These conditions were derived from laboratory experiments carried out primarily in the Western Regional Research Division at Albany, California of the US Department of Agriculture in the 1950s. A thin layer of paddy was dried with heated air using different grain moisture levels, quantities and rates of air, air temperature and time of exposure. Effect of each condition on grain breakage (or head-rice yield) was noted. The conclusions were that if the number of passes (with intermediate tempering after each) was increased and the amount of moisture removed in each pass was kept within strict limits, and if the time of a pass was kept low, then the grain damage could be avoided even while using a very high air temperature (>70°C). Thereby in-dryer drying time could be reduced, i.e. drying could be very fast, even while avoiding grain damage. All these results were expressed in a single diagram (Fig. 3.10) for each variety.

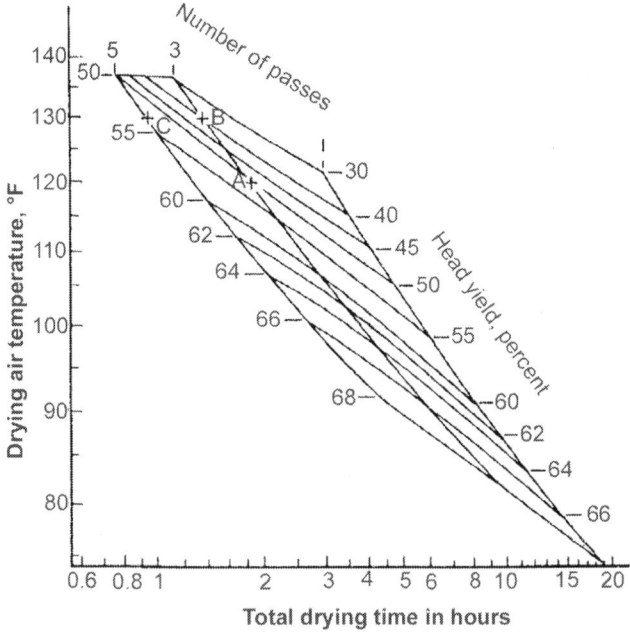

Figure 3.10 Relationship between number of passes, air temperature, drying time and head-rice yield in heated-air drying of paddy [Reproduced with permission from Wasserman et al. (1957)]

Further reading

Theory and practice of drying of grain, including paddy, is discussed in detail in several text books / chapters of books, viz.:

- Dachtler (1959)
- Araullo et al. (1976)
- Esmay et al. (1979)
- Steffe et al. (1980) – [In Luh (1980)]
- Foster (1982) – [In Christensen (1982)]
- Kunze and Calderwood (2004) – [In Champagne (2004)]

The subject of grain damage (cracking) during heated-air drying of paddy has been discussed in detail in:

- Kunze and Calderwood (2004) – [in Champagne (2004)]
- Bhattacharya (2011)

Storage of paddy and rice

4.1   Introduction

The importance of food grains, especially cereals, in sustaining our society has been referred to before. Their importance lies mainly in the fact that these grains are 'dry' and hence can be stored for long periods. Hence their suitability to serve as reserve staple food is so valuable to human civilisation.

While the above statement is true in a general sense, it is not to be taken literally. Cereals are undoubtedly a storable commodity, but primarily so in a relative sense; i.e., in comparison to perishable food. If perishable food cannot be stored for more than a few hours or days, cereals can undoubtedly be stored at ease for days or weeks or even months. But that does not mean that they can be stored at will without serious loss or deterioration under all conditions.

Cereal grains are actually seeds with an ample provision of reserve food for supporting the germination and initial establishment of the embryo when the time comes. That is the seed's destiny. However, human beings have so manipulated the agronomy and genetic material by appropriate breeding that the crop can be produced in abundance. So a tiny part of it can be used as seed and the lion's share can be used as human food. This is just like the case of milk, its biological objective being to nurture the calf, but the bulk being appropriated by humans for their own purpose. But precisely here the problem arises that other life forms demand a share of this surplus booty. This too is in the natural scheme of things. Humans must then either agree to share, or else so store the food as to protect it from these unwanted predators.

These predators are of four or five kinds. These are, in an ascending order of organisation, fungi or moulds, insects, birds, rodents and unauthorised humans. Our task is then is to so arrange the storage conditions and storage system as to prevent these predators from laying a hand on the grains, or at any rate to minimise it.

Birds and rodents can be prevented from access to the grain by creating appropriate structural barriers, along with selective use of repellents, baits, traps and poison if needed. But moulds and insects cannot be prevented from invading the grains by structural barriers alone or by traps or even poisons alone. For that, one must study (a) the properties of the grain in mass, (b) those of the environment, i.e. the atmosphere in general as well as what surrounds the grain, and (c) the behaviour and life cycle of the insects and moulds. Such study would enable us to locate the strong points of the grains and the surroundings that we can use to our advantage, and also the weak points of the predators which we can use to weaken or eliminate them. So we can achieve our goal of storing the grain for long periods with minimal or no deterioration or loss.

4.2   Grain characteristics which affect its storability

While considering the properties of grain, rice in our context, there are two types of properties that can potentially affect its storability: physical and biological.

4.2.1   Physical properties

There are at least four physical properties of grain that impact on its long-term storage. First, an important point is that paddy (as well as milled rice), in fact all particulate matter, forms a heap whenever placed on a plane. Rice, like all particulate matter, has internal grain-to-grain and grain-to-surface friction. This friction prevents grain from flowing like a liquid and makes it form a heap in the form of a low cone. The angle of the cone to the horizontal is called the angle of repose, which is a characteristic of all individual grains. The angle of repose of both paddy and milled rice is approximately 36°.

This frictional property has two important effects. One is that the grain does not flow like a fluid. The weight of the grain mass is largely borne by the walls of the storage structure and does not fall on the bottom. So if the grain is stored in a container and an opening is made somewhere towards the bottom, the grain would not just flow out. It would do so only according to some characteristic rules dictated by its frictional properties. So these behaviours have to be kept in mind while building the storage structures as well as arranging the handling of the stored grain.

The other effect of friction, of special importance in its storage, is that nearly half the space in a volume of the grain is actually empty space. This empty space is caused by the same grain to grain friction and is called the porosity, i.e. the proportion of the total volume occupied by the empty space. Paddy has a porosity of about 50% and milled rice about 40%. This porosity gives rise to a number of consequences which directly or indirectly impacts on the grain's storage. The empty space harbours pests, provides air (oxygen) for the grain as well as pests to breathe, causes the internal air to generate convection currents if there is difference in temperature between two parts in the mass, creates partial resistance to air flow if one tries to force air into the mass, and creates an atmosphere for the grain to exchange humidity (water vapour) with.

The second physical property that impacts on the grain's storage is that grains including paddy and rice are excellent insulators preventing conduction or dissipation of heat. This property causes different portions in a mass of grain to retain different temperatures, if present initially or suddenly developed, for relatively long periods. Such temperature differences in turn cause potential convection currents with consequent loss and gain in moisture in different portions of the mass. Such translocation of moisture from one part to another in a big mass, caused by temperature difference, is potentially hazardous. It creates opportunities for increase of moisture content in small portions of the bulk which would promote fungal activity there and its further consequences.

A third important physical property of grain relevant to its storage is its hygroscopic nature. As already explained in Chapter 3, grain like all biological material is hygroscopic, i.e. it exchanges moisture with its surroundings. It gains moisture from the surrounding air whenever the vapour pressure in the air is higher and loses moisture to the air whenever its own vapour pressure is higher. This property has great implication in terms of storage of grain just as it impacts on other actions such as drying or trade. This will be referred to further while discussing the impact of environment on storability.

A fourth physical property relevant in bulk storage of grain is what is called spoutline. This happens especially when the grain is not cleaned well and includes excessive amounts of dust, fines, impurities as well as broken and damaged grain. Whenever such grain is poured into a bin or storage structure, fines accumulate at the top centre. Thus they fill the inter-grain space there, while the good whole grain tends to slide down the slope into the periphery. As a result, a vertical core forms extending from a little above the bottom to the peak. The width of the core depends on the width of the bin and the amount of impurities. But the core is a solid mass. It prevents air circulation, accumulates heat and promotes activities of insects, moulds and other organisms. If left unattended, a large spoutline invariably leads to heating, moulding and serious spoilage of the mass.

4.2.2   Biological properties

Grains also have biological properties that impact on their storage. The main point to note is that the grain is a living thing and it breathes. Respiration is a process involving air. It uses oxygen and releases carbon dioxide and uses up dry matter. The most important thing to note is that it produces moisture and heat and thus recreates the process in a self-perpetuating cycle. The chemical equation for aerobic respiration, assuming glucose as substrate, is

$$C_6H_{12}O_6 + 6\,O_2 \rightarrow 6\,CO_2 + 6\,H_2O + 673\ \text{kcal}$$

This respiration is promoted both by temperature and moisture, the latter being more important. A convenient way to measure the rate of respiration is to measure the evolution of carbon dioxide. The rate of evolution of carbon dioxide nearly doubles with each increase of 10°C in temperature. But the effect of moisture is more prominent. At 38°C and 13% moisture, the amount of CO_2 released per 100 g of rice is approximately 0.5 mg per day. But at 17% moisture with the same temperature, the amount of CO_2 now released is 18 mg (Fig. 4.1). This fact shows the pronounced effect of moisture and also the self-accelerating nature of the process. For the products of respiration (heat and moisture) are both precisely those that promote further their respiration. One should note here that it is not only the grain but also the associated insects and moulds that respire in the same manner.

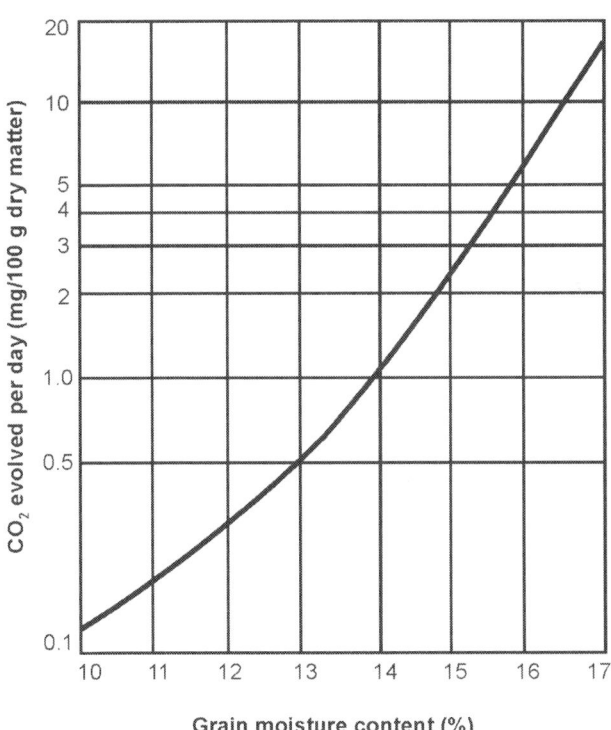

Figure 4.1 Respiration rate of paddy with increasing moisture content (at 37.8° C). Reproduced from Bailey (1940).

4.3   Properties of the environment

The environment can be considered from two standpoints. One is the overall environment outside and the other the microenvironment within the mass.

4.3.1   General climate

The former refers to the general climate of the geographical location. This can be an important consideration, for the broad approach to the storage system would depend on the overall weather pattern of the location. The fundamental point to note is whether the location is situated in the tropics or in the temperate zone. The overall weather pattern in the tropics can be considered to have an average value of close to 30°C and 85% RH. These conditions are highly conducive to the survival and growth of pests and therefore call for specific procedures and precautions to be adopted for storage of grain. Specifically, special steps have to be adopted to prevent fungal and insect infestation. There are other specific differences between temperate and tropical procedures. As an example, one can note that the prevailing low ambient temperature in temperate countries (often 10°C or lower for long periods) can itself be taken advantage of to aerate the grain under storage in bulk. This itself can work as a preservative or as a deterrent to proliferation of pests. Such luxury would hardly be available in the tropics or subtropics. Another example is the situation with respect to the sunlight. The sun especially in summer can be so hot in the tropics that it can heat the relevant part in the storage structure to a very high temperature. This would generate convection current and its consequences within the grain mass inside. So a storage structure and a material of construction that are suitable for temperate climate may not be suitable at all for a tropical area. In other words, the storage system cannot be considered in isolation of the latitudinal location of the place.

Similarly, the overall technological status of the country in question as well as the general scale of operation are other relevant factors. What is suitable or prevalent in an industrialised country in the temperate region may be unsuitable in a developing country situated in the tropics. One should not forget in this context that the lion's share of the rice crop in the world is grown in the developing countries situated in the tropics and subtropics; all the more reason why this factor assumes special importance in the context of rice storage.

4.3.2   Microenvironment within the grain mass

The second point in relation to the environment is the microenvironment surrounding the grain. In other words, it refers to the interstitial atmosphere within the grain mass that surrounds the individual grains. This atmosphere in the empty space within the grain mass is in constant interaction with the grain especially in relation to moisture. The properties of equilibrium moisture content (EMC) of the grain and equilibrium relative humidity (ERH) of the surrounding air (Chapter 3) come into the picture here. Grain is hygroscopic and constantly exchanges its moisture with the surroundings. It gains moisture if the vapour pressure of the surrounding air is high, and conversely loses moisture if its own pressure is higher. The role of relative humidity (RH) comes into play here. Humidity can be expressed in two ways. One is the absolute humidity which is expressed in terms of grams of water vapour in a unit volume (1 cubic metre) of air. The other is in terms of relative humidity, i.e. the absolute humidity expressed as a proportion of the saturation absolute humidity of air at the given temperature. Saturation humidity of air rises exponentially with increasing temperature (Fig. 4.2). Therefore, the relative humidity value of a given air falls precipitately as the temperature of the air increases at any given value of its absolute humidity. It is the relative humidity which is important in relation to the hygroscopicity of biological materials including grain.

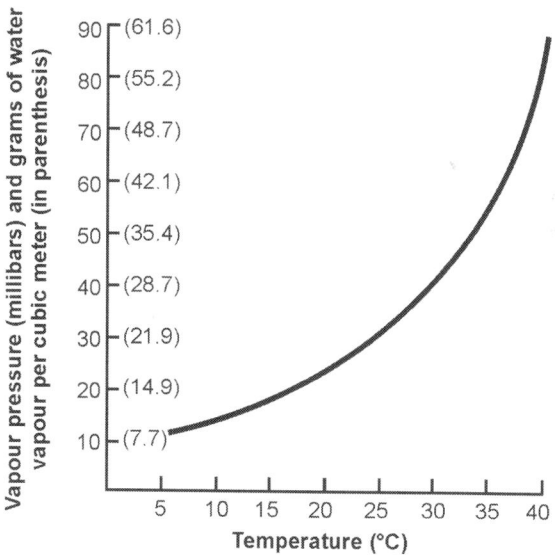

Figure 4.2. Saturation vapour pressure of air at different temperatures

The main point of importance here is that, of the two main infesting predators of grain, viz. insects and moulds, proliferation of mould is primarily related to the prevailing RH. There is a cut-off RH value below which moulds cannot remain active and above which become increasingly active. Thus mould infestation of grain is primarily related to the relative humidity of the atmosphere surrounding the grain or in turn to the grain moisture, which two are interrelated. The hygroscopic nature of grain and the relative humidity of the atmosphere thus play a crucial role in storage of food grain.

Another point to understand in this context is that the average moisture of a consignment of grain is not necessarily a valid index of its safe storability. The average moisture content is the appropriate index for determining the cost or pricing and for settlement of claims in trade. But it need not necessarily have a relation to the safety of storage of the consignment. If any portion of the grain has a higher moisture content, or if two lots with different moisture contents are mixed together, or if moisture content in a small area suddenly increases due to pest activity, the resulting increase in the relative humidity in the area can easily set off a chain of events causing serious damage to the whole lot. Similarly, local temperature fluctuation, caused for example by exposure to sun in one area, may set off a train of events in an area including generation of convection currents and deposition of moisture in a cooler area with consequent proliferation of pests there. The hygroscopicity of grain and its interaction with the surrounding air thus play a crucial role in the storage behaviour of grain.

There is another interesting point in relation to the importance of the surrounding atmosphere in storage. If the structure of the grain store is such that it can be sealed, the scenario may play out in another style. This happens normally during underground storage. In olden days, grain often used to be stored in underground pits with the access mouth sealed off. What happens in such conditions is that the oxygen is gradually used up by biological activity of the grain and the pests together, while the concentration of carbon dioxide increases ultimately to toxic proportions. As a result, a time eventually comes when all biological activity (of grain or pest) ceases. Then any further deterioration is stopped and the bulk of the grain remains in good condition. Although this practice is rarely followed now, one should take a note of the phenomenon from the stand point of general principles.

4.4   Behaviour of moulds

Infestation by fungi or mould is a serious cause of damage or loss or deterioration in quality of grain in storage. Spores of moulds are ubiquitous and present everywhere. They are present in the crop field, in transit vehicles, in containers, around and within the storage structure and in the air. Their number can be reduced by appropriate cleaning and maintaining hygiene, but their access cannot be entirely stopped. So protecting grain from destruction by moulds requires an understanding of their life style and creating circumstances inimical to their growth.

Moulds are of two kinds: field fungi and storage fungi. Field fungi invade grains in the field under appropriate conditions and cause deterioration mainly in the form of pigment spots, damage to the germ, shrivelling, etc. But these field fungi do not generally survive for long once the grain is dried after harvest and gradually die out during storage. It is the storage fungi that take over during storage of the grain if circumstances are favourable. The storage fungi are of a number of species generally belonging to the geniuses *Aspergillus* and *Pencillium*. They cause discolouration and pigmentation, damage or destruction of the germ, loss of solid matter, caking, moulding (forming a webby mass with grain) and in general severe loss in extreme cases. There are also certain mould species which form and release toxic chemicals, such as aflatoxin. Although instances of formation of phytotoxin are rather rare in the case of rice and paddy, the theoretical possibility is always there.

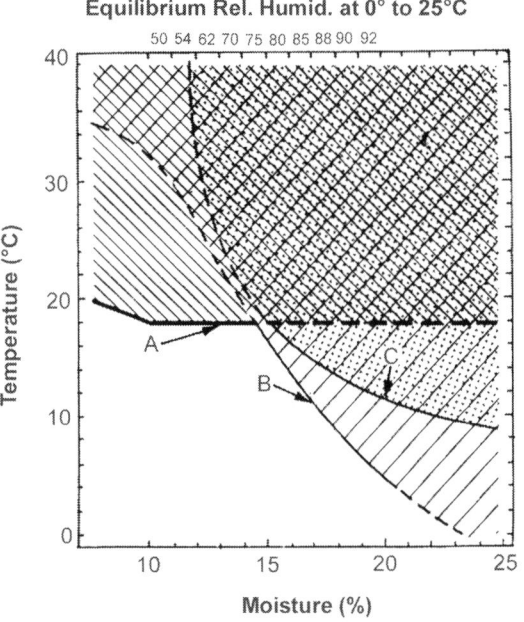

Figure 4.3 Levels of temperature, moisture content (and relative humidity) for insect and fungal heating. A: lower limit for insect heating. B: Lower limit for germination C: lower limit for fungal heating [Reproduced from Esmay et al. (1979)]

There are two significant points in relation to the life style of moulds from our context. One is the relation of the mould to the prevailing relative humidity. Moulds generally have very precise requirements of relative humidity (or grain moisture, which are indeed interrelated) for their survival and proliferation. By and large, there are very few moulds that can remain active in an atmosphere of less than 70% RH. They can survive in such atmosphere but cannot remain active. One should note that their spore can survive even under relatively hostile atmospheres. Clearly the chief route to control mould infestation is to maintain appropriate grain moisture such that they cannot be active (see Fig. 4.3). Ensuring that the grain moisture is maintained at 13% moisture or below (at the most 14% moisture or even 15% moisture for brief periods) and preventing absorption of moisture during storage are thus the surest ways to protect the grain from fungal infestation. However, as stated earlier, this refers to the moisture content of all the grains in the mass and not the average moisture of the lot.

No doubt temperature also has a role to play. But that is of less importance. Moulds have a much wider tolerance towards temperature. They generally become quite inactive below 10°C and are not at all comfortable above 40°C (Fig. 4.3). But they have a fairly wide tolerance within these limits, although their optimum temperature for growth is approximately 27°C. But their requirement of an appropriate relative humidity, as mentioned, is very strong and it is regardless of the prevailing temperature. In an atmosphere with relative humidity below their specific requirement (generally below 70% RH), their vegetative and sexual development is totally stopped even though they could survive under these conditions.

One should further note that the very act of proliferation of fungal activity is self-accelerating. Any fungal activity would necessarily generate moisture and heat, which in turn further promote precisely the same activity. This point is especially relevant to development of what is called a 'hot spot'. Due to any uneven distribution of moisture within a given mass of grain or due to excessive contamination in a small part, there may be a small area in a mass where moulds start to proliferate. The resulting generation of heat and moisture and accelerating pest activity, combined with the poor conductivity of grain, can create a dangerous focus ('hot spot') that may itself become a trigger for spreading of the infestation and could even lead to eventual total destruction of the stock.

It may be noted that theoretically speaking there are ways to control fungal proliferation with the use of chemicals. Propionic acid in particular and acetic acid have been shown to be capable of controlling mould growth. The reduced pH does not generally support proliferation of mould. However, nonvolatile chemicals do not evaporate away. Therefore, there could be serious problems with such use in practice, and this route is rarely used for fungal control except in an emergency (e.g. prestorage temporary preservation of damp grain).

4.5   Behaviour of insects

The other serious pest of rice is insects. In fact, considering tropical countries, which is what particularly relevant for rice, insects are serious pests of the grain. Fortunately paddy is not so susceptible to attack by insects, for the husk provides good protection. But that does not mean it is totally immune and safe, only relatively so. But milled rice is highly susceptible to attack by insects.

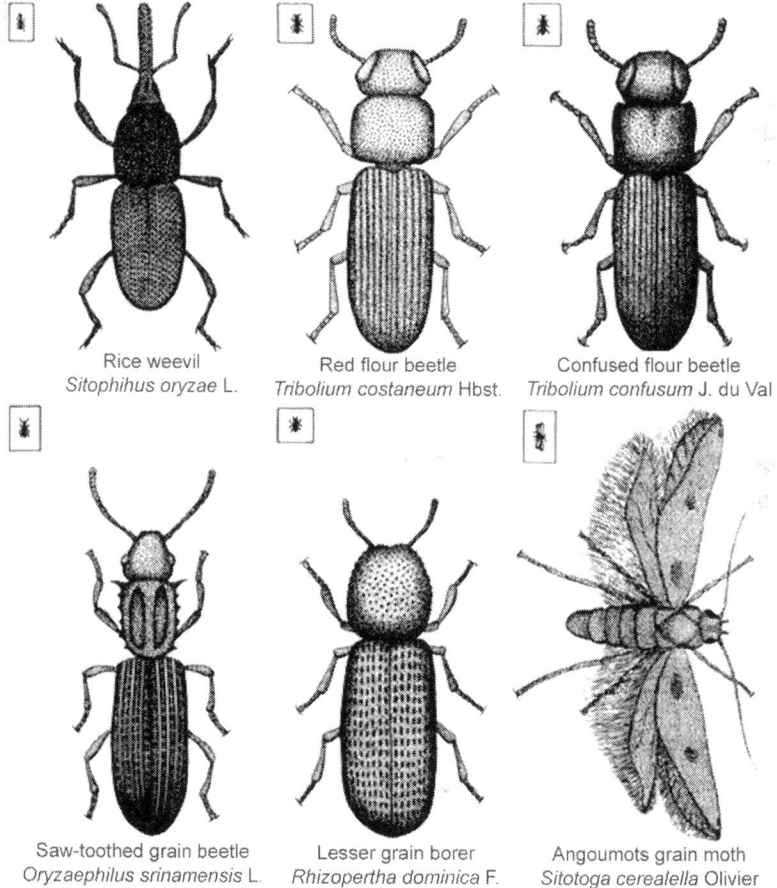

Rice weevil
Sitophihus oryzae L.

Red flour beetle
Tribolium costaneum Hbst.

Confused flour beetle
Tribolium confusum J. du Val

Saw-toothed grain beetle
Oryzaephilus srinamensis L.

Lesser grain borer
Rhizopertha dominica F.

Angoumots grain moth
Sitotoga cerealella Olivier

Figure 4.4 Some insect pests of stored paddy and milled rice
[Reproduced from Esmay et al. (1979)]

Insects attacking rice are many. The more important among them are the beetles *Sitophilus oryzae* (rice weevil), *Rhizopertha dominica* (lesser grain borer), *Tribolium castaneum* (flour beetle) and *Trogoderma granarium* (Khapra beetle), and the moth *Sitotryoga cerealla* (angoumois grain moth). There are several others but these are less common. Pictures of some of the major rice insect pests are shown in Fig. 4.4. They all have the usual life cycle and go through four discrete stages: egg, pupae, larva and adult. An adult beetle, which is a voracious eater, has a body having three distinct parts: the head, the thorax and the abdomen, with three pairs of legs. Under optimum temperature and humidity conditions, their life cycle is quite short, about 4–5 months. The female lays about a hundred eggs at a time, which hatch in about a month's time. It is clear that under optimal conditions, they multiply at great speed, the number going up to 10^8 in a span of about 6 months. Their spores are ubiquitous and present everywhere. So contamination takes place from the field, in the bag or the container, during transport or in the storage structure. Proper cleaning of the grain as well as maintenance of good hygiene are essential to keep the contamination to the minimum.

As for their temperature and humidity requirements, two points are to be noted. Insects are strongly sensitive to temperature. They become totally dormant below approximately 18°C and cannot survive beyond 41°C (Fig. 4.3). So their operating temperature requirement is within the range of about 19–40°C with an optimum of about 29°C. This requirement shows that insect infestation of grain can be controlled relatively easily by temperature alone in the temperate climate. In these countries, air temperature can often come down to much below this range. Therefore, by the simple act of aeration of the grain in bulk storage at such periods, infestation by insects can be kept under control. Unfortunately, the situation is by and large quite the opposite in countries in the tropical region, which is what all the more relevant in the case of rice and paddy.

Unfortunately, control of humidity too is not very helpful either. Insects are not so sensitive to the relative humidity of their surroundings. Although they may not be quite very active in a dry atmosphere (65% RH or below), they are reasonably tolerable to any humidity within their comfort zone of about 20–40°C temperature. This is doubly unfortunate, for the prevailing temperature in the tropics falls precisely in this range without much help from the humidity either.

Insects and moulds generally complement each other and flourish hand in hand together. Insects themselves harbour fungal spores and the life activity of each promotes the other. When insects and moulds become active, the products of their respiration and life activity, viz. moisture and heat, help to promote further respiration and further activity in a vicious cycle.

It is apparent that control of insect infestation of paddy primarily requires total attention to maintenance of proper hygiene and cleanliness. Apart from that, maintaining a low temperature whenever feasible is most important. While this is hardly achievable in the heart of tropics, it is not impossible to be done at the higher latitudes, especially in the subtropical regions, which is also important in terms of paddy crop. Control of relative humidity, in any case difficult to do in the tropics and subtropics, may not be crucial but can still be helpful. For one thing, a low humidity may itself render the insects rather lethargic. But more importantly, it would keep the moulds in total check and thus remove one contributing factor.

One can see that one may have to occasionally fall back on the use of insecticides to keep inspects under check in the tropics. Insecticides used to be widely employed for disinfestation of grain decades back but their use is now generally frowned upon. Insecticides have many hazards and limitations. These do not necessarily act on all genuses and species and in all stages in their life cycle. These may also have more harmful effects on the parasites of the insects and thus indirectly promote the infestation. And these may have mammalian toxicity. So these have to be used with discretion. Moreover, insecticides can leave residues in the grain in quantities that may not be permissible or they may have long-term ill effects on humans. So several insecticides that were being used in grain storage have now been banned one by one. The more or less lone insecticide that is permissible for use as of now (2014 CE) is aluminium phosphide, which releases the poisonous gas phosphine when in contact with water or humid air. But even its indiscriminate use is frowned upon. Others (such as malathion, methyl bromide, etc.) are not permitted to be used directly on grain, but can still be used to disinfest walls of structures and the surroundings of the premises. Therefore, adoption of appropriate storage practices to prevent further cross infestation, after one disinfestation with fumigation, is essential.

4.6   Storage structure

Obviously grain has to be stored in a shelter or a building or some sort of a structure. The size will depend on the amount of grain that is to be stored. In ancient times, much of the aggregate surplus grain of the society was owned and held by numerous peasant owners. So by and large only a small quantity would have had to be stored by each. Containers or structures for storing these amounts would necessarily have been small. As society progressed and organisation (urbanisation) developed, more and more concentrated or centralised storage had to be done and bigger and bigger structures were called for. This is a process of transition that is, at one extreme, still in the stage of beginning in some countries. It has reached a very high level of centralisation in some, at the other extreme. Not only the size but the style and material of construction of the structure would also depend on where exactly in this chain of transition the particular place is located.

In the case of traditional small holdings, not only the size would be relatively small, the material of construction of the structure would also naturally be restricted to locally available indigenous and organic materials, such as reeds, bamboo, plant stalks, wood and mud. Although best efforts may be made, these structures may or may not be quite effective for safe storage of grain in the humid tropical climate. These may undoubtedly protect the grain from rain and overt exposure. But these would not necessarily be water-vapour tight, insect resistant and rodent proof. In the high-temperature and high-humidity atmosphere of the tropics, especially during the rainy seasons, these structures may not be good enough to prevent entry of humidity and pests and rodents. Structures and vessels made of baked clay (ceramic) are far superior in this respect. Despite these drawbacks, however, one should remember that these structures are more successful to protect from temperature fluctuation than many modern structures.

It is quite possible to improve and upgrade these traditional structures. Mud-wall structures can be made humidity-tight by incorporating plastic or bituminous material or even cement during construction or as a lining. Good plastic films are very successful in providing air or water tightness. Metal bins are obviously the best in this respect for providing protection from access to humidity or insects and other predators. However, these bins are susceptible to temperature variation if placed in the open. Indoor storing of clean and dry grain in metal bins is obviously most successful in protecting grain from not only moisture and predators but also from temperature fluctuation. Providing a metal-sheet protection for the bottom one metre can make these bins rodent proof.

Systems of large-scale storage of grain in modern times have evolved out of these basic traditional systems. However, one fundamental point in modern storage is, the precise system will depend on whether the proposed storage is in bags or in bulk. This decision depends upon the economical and existential factors. The prevailing system in the society in question, the existing transportation facilities, the availability of handling systems are all relevant to the issue. If bulk-storage facilities are created in a place where bringing in of grain and taking it out have to be provided in bags, it may not turn out either efficient or economical. If the existing infrastructure in a society is not oriented to bulk storage, it may be prudent to restrict bulk storage to only some certain specified needs, such as operational silos or long-term holding. One should note that bag storage is labour intensive, while bulk storage is capital intensive. So the scale of operation, the cost of labour and the availability of capital are also all relevant factors in the issue. Another point is that the cost of a building for bag storage is much lower than that for bulk storage including the necessary handling facilities. On the other hand, a bulk-storage system uses the available space almost to the full, while a substantial proportion of space (perhaps about 30%) has to be necessarily left vacant when storing in bags. This open space is required not only for bringing in and taking out of the bags as well as for moving around for inspection and other operations, but also for leaving a clearance between the bags and the walls to protect them from temperature fluctuation and dampness. Again bag-storage structures are relatively more difficult to make vapour tight or air tight, which is inbuilt in structures for bulk storage. In bag-storage systems, bags provide a relatively much more surface area for loss of excess moisture but also for absorption of moisture if the structure is not air tight and the weather is humid. Similarly, hot spots do not easily develop in bag storage, while it is a constant hazard in bulk storage.

The storage structure should have the following capabilities: (i) The structure should be water tight to prevent liquid water from entering and wetting the stored grain. (ii) It should be air tight to prevent the absorption of water vapour from outside atmosphere having high humidity. (iii) Alternatively, or optionally, it should be provided with arrangement for circulation of air throughout the mass. (iv) It should be capable of being totally sealed such that the grain may be fumigated if needed. (v) The building should be rodent proof and bird proof. (vi) Finally it should be economically designed.

Bag-storage buildings and warehouses should have water-proof roof, walls and floor. It should also have openings meant for controlled ventilation as well as full-building fumigation. But these openings should be capable of being entirely sealed to prevent water vapour from being absorbed whenever the atmosphere is humid. At the same time, these openings will permit natural or artificial ventilation when the atmosphere so permits. Alternatively, groups of bag stacks can be fumigated under suitable plastic covers. The roofs of such warehouses are often made of metal or corrugated sheets, which can be the worst possible choice unless insulated. In tropical countries, they can cause wide temperature fluctuations with its consequences. Such roofs should either be adequately insulated with insulation blocks or provided with suitably built false ceiling. The attic above such false ceiling should be provided with either natural or artificial ventilation system for adequate control of temperature. One needs hardly to say that the building should be rodent and bird proof. The building should be built on a raised plinth so as to make it rodent proof. Doors and openings should have proper plastic hangings to prevent entry of birds. Since aeration is not very convenient or successful in such warehouses, only properly cleaned and dried grain should be put in for storage. Building the walls with blocks, concrete or brick and masonry can provide protection against liquid water and humidity and also a fair amount of side-wall insulation. Providing wooden planking above the masonry floor for placing the bags is mandatory to prevent absorption of moisture as well as cold by the bags from the ground.

Bulk-storage systems must also meet the basic requirements of being water proof, vapour proof, as well as rodent, bird and insect proof. Side wall insulation, if possible, is extremely helpful because the grain here is in direct contact with the wall. Metallic walls and roof are not the best, especially for tropical and subtropical countries. As a matter of fact, metal structures were developed for temperate countries and are not suited in the tropics. Metal walls facilitate both inward and outward transfer of heat. Besides, metallic walls absorb radiation as per the reflectivity of the external surface and its re-radiation capability of long-wavelength (low-temperature) heat energy. Aluminium and polished galvanised steel can reflect radiation well but cannot reradiate long-wavelength energy. This can be partly taken care of by painting the walls and roofs with good quality white paint which is a good reflector and radiator.

These structures for bulk storage can be small or big, flat or upright, with flat bottom or hopper bottom (silo). The structures made of reinforced concrete are best but are much more expensive.

4.6.1   CAP storage

One ingenious system for bag storage of grain (paddy, not milled rice) has been developed in India, especially by the efforts of the public sector organisation, the Food Corporation of India. This is called CAP storage, i.e. cover and plinth storage. In this system, a slightly raised cement masonry floor is provided (the plinth). Bags of grain are arranged (on wooden planks at bottom) in a semi-pyramidal form to form a tip line on the top along the long axis. The two slanting roofs thus formed on top are covered with heavy plastic and tarpaulin. Even though left in the open under rain and shine, this system has been quite successful for reasonably safe storage of grain for months on end. This is especially true of North India after the winter harvest when the weather is cool and dry. The cover is rolled up for hours every day on fine weather days, thus allowing the slightly damp grain to gradually dry on its own. However, CAP storage is used only to store paddy. Milled rice is never stored by this system. This innovative system, though rather high in labour cost, costs next to nothing in capital, is very flexible and quite successful. Its one main drawback is that it is not rodent proof and extra efforts have always to be made to keep rodents under check. Apart from traps and poison baits, fumigating the surroundings with malathion or lindane can act as a repellent for rodents.

4.7   Moisture migration

Translocation of moisture from one part to another is a hazard in grain under storage particularly in bulk-storage systems. The grain that is certified to have an average moisture content within the limits for safe storage may not be really safe for reasons explained earlier. But even grain that has been fully dried properly to a safe storage level (e.g. 13% moisture) may not necessarily stay in good condition for long without certain precautions. Apart from direct and overt infestation by moulds and insects, moisture migration is another important hazard in bulk storage which may indirectly cause infestation, heating and consequent deterioration.

Moisture migration may happen due to many reasons, viz.

(a)   grains having different moisture contents have been mixed together,

(b)   cleaned and uncleaned grain have been put together,

(c)   warm (or cold) grain is added on cold (or warm) stored grain, and

(d)   the storage structure is exposed to seasonal and diurnal (daily) temperature fluctuation.

The reason of moisture migration for the first three factors is clear enough. Higher moisture grain will necessarily create a surrounding atmosphere having a higher RH and promote pest activity. Similarly uncleaned grain carrying extra load of eggs and spores of pests and moulds may trigger infestation activities. This may promote local biological activity and consequent localised heating which, facilitated by the well-known non-conducting nature of grain, may cause 'hot spots' to develop. The resulting convection current would carry moisture vapour from one portion to another. It should be noted that it is not only increased humidity in interstitial air that would carry moisture from one place to another. Warmer air even at the same RH carries more water vapour than cooler air and would deposit the extra moisture on the cooler grain when it is cooled.

However, an unsuspected culprit with respect to moisture migration is temperature fluctuation within the mass. If storage structures are not insulated from the climatic environment, the grain temperature may vary in different portions. Seasonal climatic changes and diurnal variation in solar radiation cause the grain next to the outer surface to either gain or lose heat and attain a temperature that is different from the rest. As air within the mass is warmed, it expands and then rises because of its lower density. Alternatively, cooler air becomes heavier and tends to move down. As the warm air (with lowered RH) moves up, it can pick up some moisture from the nearby grain. But when the warm air cools after coming in contact with a cooler area, it would tend to release the extra moisture on the cold area. This additional moisture in a localised area of stored mass of grain would again set in the well-known chain of events. Respiration as well as other activities of moulds and insects would increase and self-accelerate. These developments again create 'hot spots'. Besides, there may also be condensation near the bottom or under the roof if these areas get cooled by the weather outside, leading to increased biological activity and damage. The intense daytime solar radiation leading to external heating in tropical countries can easily cause serious convection currents in the stored mass leading to events described. This is then one of the ubiquitous hazards of large-scale bulk storage in tropical countries.

Conversely, if the grain mass can be maintained at a stable temperature, potential for air convection within the mass would be automatically reduced with corresponding reduction in storage hazard. This is achieved either in smaller structures by placing them inside the home or in a sheltered area, or else by providing for aeration of the grain mass in a big structure.

One way to minimise or reduce temperature fluctuation in the grain due to diurnal variation of temperature is to orient the storage structure correctly. It is the eastern and western walls in the tropics that get the bulk of the solar heat. Therefore, any rectangular storage system should be so oriented that its long axis is oriented in the east-west direction.

4.8   Aeration

Grain stored in bulk in large quantities for any substantial length of time (e.g. for more than a few months) usually needs to be aerated at suitable intervals. Aeration means moving a relatively low volume of air through the mass of grain. Its purpose is not to dry the grain, even though a small amount of drying may occur. Its main purpose is to equalise temperature and to break 'hot spots' in the mass.

Bulk storage of grain in large structures came gradually into vogue in Europe and USA after World War II during the time of grain surplus. Consequently, as a corollary, aeration of bulk stored grain too gradually came into use, especially with the research investigations carried out by the US Department of Agriculture in the 1950s. Initially grain used to be moved from one filled bin to an empty bin for the same purpose. This process was called 'turning', the object being to break up the mass and mix. This used to cause a certain amount of grain damage and also energy cost. So gradually this was replaced by aeration, for it did the same job but cost less and caused no grain damage. Both processes achieved the same purpose. In turning, grain is moved through air. In aeration, air is moved through grain.

The purposes of aeration are the following:

- To reduce the temperature of the stored grain whenever the grain is warmer and the atmosphere is cooler, especially when the atmospheric temperature is less than 18°C. This objective is generally quite feasible to be achieved in temperate climate, and perhaps also in subtropical climate on occasions. But it is hardly achievable in tropical climate. As explained under the sections discussing mould and insect infestation, maintenance of such a low temperature can itself help in providing safe storage by preventing or retarding pest activity.

- To prevent moisture migration. Moisture migration caused by any of the factors described above may itself be a source of pest activity and quality loss in grain. Aeration is an effective method of equalising temperature and preventing convection current.

- To equalise grain moisture. When different lots of grain having different moisture levels are bulked together, aeration can help to expedite the equalisation of the moisture within the mass.

- To remove odours from grain. Mouldy or musty odours, chemical odours from fumigation, sour or fermented odour from organic acid preservatives, can be removed or at least diminished by proper aeration.

- To help in the application of fumigation. Aeration may help in application of fumigants and in their better distribution.

- To remove dryer heat. The grain coming out of heated-air dryers in the last pass may be quite warm, which in itself may be a cause of promoting deterioration. Aeration in such cases can help in dissipating the high temperature.

- To help in temporary safe holding of moist grain. In modern mechanised agricultural systems, usually large amounts of moist grain suddenly arrive at a time in drying centres. Holding a part of it temporarily in safe storage for short periods, pending drying, may become incumbent. It has been found that providing aeration can help such moist grain to be held undamaged for short periods. Even partially dried grain can be held thus for some time to ease the pressure on the dryer.

As mentioned above, aeration involves moving a relatively small volume of ambient air through a mass of grain. The intention is not to dry the grain but only to cool it or equalise the temperature. As such, the air volume needs to be small, usually in the range of 0.07–0.15 cubic metre of air per minute per tonne of grain (1/20–1/10 cubic feet of air per minute per bushel of grain). Obviously, as the grain is in bulk and in a large quantity, appropriate systems for air distribution throughout the mass have to be provided for in the storage structure. Appropriate ducting and/or a false flooring system have to be provided in the structure for the passage of air, as was shown in Fig. 3.7 in Chapter 3 (Drying of paddy). As air would normally tend to take the path of least resistance, the air duct system has to be carefully planned to equalise the distance of air travel in all directions. Also in a big bin, the air has to pass through a substantial depth of grain, so the air must have the ability to penetrate the mass. In other words, the operating fan must have the ability to create appropriate static pressure. This static pressure requirement increases exponentially with increasing volume of air and also the depth of the grain mass.

One should at the same time remember that while aeration may help in better preservation, it may do the opposite if indiscriminately operated. The ambient air may be cool, but if it has a high humidity, then along with cooling, the grain may absorb some moisture. Therefore, it is necessary to ensure that the aeration system is used not simply when the air is cool but also when its relative humidity is sufficiently low (e.g. less than 70% RH).

Aeration can be useful in tropical climate as well. Even though the ambient temperature may not be quite low, aeration may still be useful. For example, even in tropical countries in summer or monsoon months, there may be several hours around noon when the atmospheric RH is 70% or lower. Aeration (or even ambient air drying) can be carried out at such times. It can at least help in equalising temperature and in breaking up 'hot spots'. Clearly, it is necessary to keep a track of the atmospheric temperature and relative humidity and to operate the aeration system only when the conditions are appropriate.

4.9   Protection against birds and rodents

Protecting grain from birds and rodents is relatively easy. The main point is to create barriers against their attack. Modern bulk-storage structures are well protected against birds and rodents. Bag-storage warehouses can be made safe from birds by ensuring that there are no openings that are not suitably screened or cannot be securely closed. Besides, operating doors should have hanging bird scarers to prevent birds from entering when operations are going along through open doors.

The same thing applies to protection from rodents. If the warehouse is built over a plinth of 1 metre in height, then it is essentially safe from rodents. However, entry ramps should be so constructed as to leave a sufficient clearance from the main plinth. Small indigenous structures should have their bottom 1 metre wall protected by metal sheets or cemented. Apart from that, appropriate traps, poison baits, fumigating burrows, etc., will keep rats in check.

4.10   Conclusion: Operational summary

Cereal grains are the major staple of humankind and hence are required for food all the year round. Clearly, they are required to be stored for fairly long periods to provide food from one season to the next. As a matter of fact, long-term storage of cereal grains is required in modern urban societies for other reasons as well, for instance:

- To provide a buffer stock of food in case of crop failure,
- To cater to the marketing and trading system, an essential part of the modern urban society, and
- To provide for seed stock required to be planted in the next season.

Adoption of informed steps, guided by accumulated knowledge of the grain as well as Nature, is necessary to ensure adequate safety of the grain during their long-term storage. The fundamental principles and knowledge, on the basis of which these action-oriented steps are to be provided for, have been discussed in the above paragraphs. The necessary steps required for safety in long-term storage of grains in general and paddy and milled rice in particular can be summarised as follows.

(1) The moisture content of the grain is the ultimate arbiter of its storability. The lower the grain moisture, the better would it store, other things being equal. Therefore, the first requirement for good storage of grain is that it should be dried under safe and appropriate conditions as early as possible after the grain has been harvested and threshed. The grain should be dried down to 13% moisture or lower for its long-term storage. Drying to 14–15% moisture is permissible for short-term temporary storage particularly if the prevailing temperature is low (15°C or lower). For long-term storage, drying down to 13% moisture or below is necessary.

(2) Impurities in the grain not only harbour spores or eggs of pests. They also cause uneven packing of the grain in the storage structure, creating resistance to air flow, preventing dissipation of heat and creating focal point of pest activity. Therefore, the next important action before storing the dried grain is to thoroughly clean it to get rid of dust, stone, mud, chaff, other seeds, etc. Thorough cleaning of the grain also helps in reducing the load of fungal spores and insect eggs.

(3) Infection by pests comes not only from the field but also, and primarily, during its handling before storage. Spores and eggs are ubiquitous and present everywhere: in the container or bag or vessel carrying grain, in the transport vehicle or carts, in the surroundings of the storage structure, inside the structure and on the implements and receptacles used for activities. Therefore, proper maintenance of hygiene before and during storage is as important as cleaning the grain. All these structures, premises, vehicles and implements should be thoroughly cleaned and fumigated before storing the grain in the new season. Old stocks of grain should be removed and if necessary stored separately. Fresh season's grain should never be mixed with the previous year's grain in stock.

(4) Apart from ensuring dryness and cleanliness, the other almost equally important condition of the grain to be stored is its temperature. The lower the temperature of the grain, the better, i.e. the safer, will it store. Maintaining a low temperature helps in keeping respiration (of both grain and pests) down and avoiding possibility of automatic heating. Besides, both insects and moulds have a strict temperature requirement for their growth and activity. This is more true of insects, although it is not unimportant for moulds either. Insects become almost totally inactive below 18°C and cannot survive above 41°C. Although not inactivated by a low temperature like the insects, moulds too are rendered less active at low temperatures. Both grow best and have their highest activity at around 25–30°C. One can clearly see that maintaining a sufficiently low temperature is by itself of great value in the safe storage of grain. Maintaining a sufficiently low temperature of 15°C or lower is not difficult in temperate climate. If the grain is not already at such temperatures after harvest, this can often be ensured by appropriate aeration. Maintaining a suitably low temperature is generally possible in subtropical climate also at least for a few months after winter harvest. But it is next to impossible to achieve in the tropical climate and in subtropical climate once summer approaches. Nonetheless, all possible opportunities of maintaining as low a temperature as possible should be taken advantage of.

(5) If insect infestation is detected after all precautions have been taken, fumigation with a permitted fumigant should be resorted to at the beginning of the storage. All other appropriate measures as detailed herein should be taken by that time, so that the need for repeat fumigation can be avoided as far as possible.

(6) It is not only necessary to maintain a low grain moisture and temperature, it is equally essential to see that different portions in a

storage lot do not have different moisture contents and/or different temperatures. Variation in temperature and moisture content in different parts of the lot would induce convection current and/or creation of 'hot spots'. These can be focal points to initiate a train of processes leading to pest activity and deterioration.

(7) The grain should be stored in a structure and system which are appropriate for the latitudinal location of the place (whether located in tropical or subtropical or temperate area). So also these should be appropriate for the socioeconomic status of the country, the organisation and the scale of operation. In particular, careful consideration should go in deciding whether to go for bag storage or bulk storage. Bulk storage should ordinarily not be considered in a place where the overall culture is oriented to bag storage.

(8) The storage structure should be as much protected from sun as possible, particularly in the tropical countries. An elongated rectangular storage structure should be oriented in the east-west direction so as to minimise solar heating. A metallic structure should be avoided if possible in the tropics. In case it is unavoidable, the structure should be painted with good white paint. RCC structures are better insulated, although more expensive. Small farmers' bins built with local organic material have better insulation. But these are less water proof. The latter can be assured by putting plastic or bituminous lining.

(9) The storage structures should be such as to be rodent proof (plinth built one metre above the ground), bird proof, leak proof (prevent entry of liquid water), vapour proof and damp proof. All steps should be taken to ensure that the grain in store is not exposed to the atmosphere at a time when relative humidity of the atmosphere is more than 70% RH. The structure should be insulated as far as possible, at least the roof if it is of metal in a bag-storage structures, and be preferably built with bricks or blocks and mortar. Bulk storage silos should be built with RCC if possible or else painted with good white paint. The structure should be such that it can be totally sealed for fumigation, but at the same time provided with facilities for ventilation (bag storage) or aeration (bulk storage). Bags should be kept only on wooden planks and not directly on the floor. Bag stacks, if necessary, can be individually fumigated under plastic cover.

(10) In the case of bulk-storage structures, adequate ducting and other arrangements should be put in place, so that the grain can be aerated whenever necessary. Aeration should be resorted to either to cool the grain or to equalise temperature or to break up the 'hot spots' within the mass.

(11) In bulk-storage structures, adequate temperature monitoring systems should be put in place before pouring in grain, so that the temperature at different locations in the mass (both at different heights and at different radiuses) can be regularly monitored. A temperature rise of even a few degrees at any spot is a sure sign of sudden increase of pest activity and is a signal for appropriate action, in particular to initiate aeration.

Further reading

Discussions on the science and technology of rice storage are available in a general way in the standard text books on rice, viz.:

- Luh (1980, 1991b)
- Juliano (1985)
- Champagne (2004)

However, detailed critical discussions on the subject are provided in:

- Araullo *et al.* (1976)
- Esmay *et al.* (1979)
- Christensesen (1982)

Reference

Bailey (1940)

Milling of rice

5.1 Milestones in the development of rice milling machinery

The rice grain, as harvested from the plant, called paddy or rough rice, is not in an edible form. The grain has an external rough and inedible covering, husk (comprising of two boat-shaped halves, viz. lemma and palea), that has to be removed first. This yields 'brown' rice, called so due to its brownish external surface. It still has some outer layers (collectively termed as bran) which do not get 'cooked' during the normal cooking process. These layers also are to be removed, partially or fully, for rice to be properly cooked. The technology for obtaining edible rice from paddy, in a form that can be cooked, is perhaps as old as agriculture itself, presumably running into prehistoric times.

The process of removing husk and the bran layers, which yields edible rice, is called rice milling. The technology presumably started with the use of mortar and pestle in its simplest form. Manual milling by hand pounding of paddy in a mortar and pestle was common for centuries. Further development of this system was the mechanised version of pestle and mortar, still operated by hand or leg. Some sort of hand pounding, in wooden or stone mortar and a wooden pestle, may still be in practice in very remote tribal areas and villages.

Use of energy for milling of rice, other than human, started some 200 years ago in South Carolina, a British colony, then in colonial America, where the cultivation of rice was introduced during 1680s. Black slaves, brought from Africa, were put to use for hand-pounding of paddy to produce milled rice. Rice was then a commercial commodity, mainly for export to Britain and other European countries. A worker, using wooden mortar and pestle, could produce from one to one-and-half bushel of rice in a day (a bushel is 45 lbs. or 20.4 kg). However, milling of paddy in mortar and pestle was not just 'pounding' of the grain, since the goal was to obtain whole, not broken or pulverised, grains. To achieve minimum breakage during the pounding process, it was generally performed in two distinct operations. During the first pounding, majority of the hull or the husk was removed, representing a relatively easy step. The second stage was polishing the rice. This was more difficult and involved detaching the bran without breakage. To achieve this required a skilled tapping and rolling motion. The accounts of that period (latter half of eighteenth century) of South Carolina mention that a skilled worker could produce 95% whole grains, while a less skilled, 'careless', or a fatigued worker could easily shatter half of the rice. Increased paddy production required more quantities to be milled. Manual labour became a limitation. Improvised pestle and mortar mills turned by farm animals were then invented in the later part of the eighteenth century. These included the 'pecker machine' in which the pestle was made to move like the stroke of a woodpecker, and a 'cog mill' in which an upright pestle was driven by a horizontal cog wheel. These machines could mill from three to six barrels a day (a barrel is 162 lbs., i.e. 73.5 kg of rice).

A water-powered (tidal-powered) rice milling system was first developed and installed in 1793 by one Jonathan Lucas on paddy farms, on the river Cooper, north of Charleston in South Carolina, USA. This was the first ever water-powered rice mill, to replace manual pounding and threshing operations by mechanical machinery in USA. Around this time, the system of flooding rice fields by action of the tide was coming into general use, but a large crop was considered a dubious blessing because of the difficulty of removing the husk from the grain. The water-mill built by Lucas was driven by a very large undershot water wheel. It was an improved tide mill with many machinery units for continuous operation. It had rolling-screen paddy cleaner, elevators, mechanical pounding system for milling and the packing system. Three persons could manage such a mill. On a favourable tide beat, the mill could produce about sixteen to twenty barrels, and larger mills up to 100 barrels of rice, per day.

It is interesting, however, to note that water-powered rice mills were also in operation on the other side of globe, in China, around the same period. A 'British Embassy', consisting of two large ships with more than 400 personnel, was sent by the King of Great Britain to the Emperor of China during 1793 which travelled through the ancient empire before returning. The account of this travel mentions, among other things, the way paddy was being cultivated and processed at that time in China. It has been mentioned that the cottage-level mill was the foot-operated, vertical pounding system (similar to that of pecker machine of USA), or a set of two circular stones between which the grain was made to pass. The large-scale rice milling was being carried out in China by water-powered mills (also similar to the water-powered mills of USA).

Jonathan Lucas subsequently installed his mills throughout the rice region of South Carolina. Figure 5.1 shows an illustration of a typical water mill existing in South Carolina, USA, during 1790s; and Fig. 5.2 shows a typical large-scale, water-powered rice mill existing around the same period in China.

Figure 5.1 Water-powered rice mill in South Carolina, USA, ca. 1800
[*Source*: Drayton 1802]

Figure 5.2 Water-powered rice mill in China during 1804 [*Source:* Wikimedia 2014]

The technology of water mills was also adopted in England for milling of paddy imported from British Colonies. In the early years of the nineteenth century, Jonathan Lucas erected an improved tide mill on the Ashley River near West Point in Charleston, South Carolina. This mill attracted a large business and made Charleston the centre of the rice-milling industry. Subsequently, in 1817, either he or his son developed and installed the first steam-powered rice mill in Charleston, which was soon replicated in the neighbouring areas. The introduction of steam power allowed the use of steam-powered pumps for irrigation in the cultivation of paddy. This assisted the development of rice production along the Mississippi river in Louisiana. Roller mills, which had been developed and were being used for processing of wheat in the northern part of USA, were being modified and adapted for processing of rice, during this period, in South Carolina and Louisiana.

An improved version of the steam-powered rice mill was set up by Jonathan Lucas III during 1840, at West Point, Charleston. This was housed in a four-story brick building. This mill, along with two others in Charleston, produced 97,240 barrels of rice in 1890. It was stated to be the largest rice mill in the USA, perhaps in the world, at that time.

The industrial revolution, which started in Europe, heralded the inventions for mechanisation and development of equipment and machinery involved in various industries. It also saw the development of new machinery in the production and processing of farm produce. As a result, the rice milling sector also had its share, and a number of machines were developed, not only for the removal of the husk and bran, but also for the connected operations like conveying, cleaning, drying and separation.

The first electrical-powered 'modernised' machines that were used for the milling of rice were the 'disc sheller' for dehusking of paddy and the 'cone polisher' for debranning of rice. These machines were first employed in rice mills of Burma (now Myanmar) around 1862. The cone polisher was an abrasive, inverted, truncated cone, coated with emery to remove the bran layers. This was developed and manufactured by Douglas & Grants of UK, in Scotland in 1860. The mechanised dehusking unit, termed as 'disc-sheller' or 'under-runner sheller', had a pair of abrasive stone discs made by wet bonding of emery, and stacked so that the emery faces were towards each other. The upper disc, with a paddy inlet port, was fixed, but the lower disc, mounted on a vertical shaft, was rotatable. Other debranning machines, with more or less similar features as that of Douglas & Grants, were developed in 1865 by Kampfnagel, and in 1895 by Schule in Germany, and in 1878 by Henry Simon in England. A mortar-type rice milling machine, wherein an iron screw was mounted on a vertical shaft within a mortar, was also developed in 1870s in USA. A horizontally oscillating, compartment-type paddy separator was patented and manufactured by Schule in Germany in 1892. An indented cylinder or trieur cylinder, developed in France in 1845 to separate seeds mixed with wheat, coffee and other grains, was inducted later into rice processing line and was employed to separate broken rice from whole milled grains.

Figure 5.3 Engelberg huller, as advertised in 1904 in USA [*Source*: Wikimedia 2014]

However, the machine which caught immense popularity for milling of rice in the traditional rice-producing and -consuming countries was the famous 'Engelberg Huller'. This was developed in Brazil by one Evaristo Conrado Engelberg, a German Brazilian mechanical engineer. The machine could remove the husks and shells from rice as well as from coffee seeds. Engelberg obtained a British patent for this machine during 1885, and a USA patent in 1888. However, a US patent for another huller machine, dedicated to rice, was granted to him in 1890. With this machine, the dehusking and debranning of the grains was achieved in a single (or preferably two) passage(s). Although it was not an efficient machine as regards its performance, it offered simplicity of operation, durability and sturdiness. The capacity of the mill (300–500 kg/hour) suited well for small-scale milling. It got, therefore, widely adopted immediately across the rice-growing countries. The rapid establishment of large-scale rice cultivation in Louisiana, towards the end of 19th century, coincided well with the debut of Engelberg huller. This led to its widespread use for large-scale milling too by deploying multiple parallel units. Initially, the huller was being manufactured in Brazil, but as the international demand increased, a branch was started in New York, USA. By 1910, the firm was not only selling the machine in USA, but was exporting it to Central America, South America, India, the Philippines, China, Japan, Russia and Africa. Among these, the highest exports were made to India and Brazil. Figure 5.3 shows the advertisement of the huller machine that was published by the company in 1904.

Alternate technologies for large-scale milling of rice with better performance were continued to be explored in Europe and USA. Technology and machinery that was being used for large-scale processing of wheat was modified and adapted for milling of rice in USA. The systems developed involved multi-stage operations for performing different jobs, connected and streamlined by handling and conveying machinery. Unit operation machinery developed in Europe for other grains were also deployed for the milling of rice. Such mills were exported to other countries too. For example, the earliest large-scale rice mill established in Queensland, Australia, was imported during 1887 from USA, manufactured by Squire & Co. of New York.

A typical large-scale American mill, operating during 1940, was a huge one, housed in a multi-story building. Paddy was lifted to the top floor and moved from one stage to the other through pipes by gravity. The intake paddy was cleaned by using scalper, cleaner and monitor machines. Dehusking was done in very large stone-sheller, consisting of an emery-coated, over-runner disc sheller, the top of which rotated and the bottom one was fixed. This was in contrast to the European design, in which the bottom one rotated. Separation of brown rice was achieved in an oscillating table separator of European design (discussed later). The debranning (whitening) was carried out in a three-stage operation. The first two were in large, tapered metallic friction polishers, about nine feet long, housed in a funnel-shaped wire-mesh cage, set in series. The last stage was carried out in an oval-shaped emery-coated 'pearling cone'. The resultant rice, which was hot, was cooled for a few hours before 'polishing' in a 'brushing' machine. This consisted of a vertical framework covered with overlapping pieces of soft moose hide or sheepskin, revolving at a high speed within a cylinder of wire screen. The machine was kept cool by blowing cool air on it. The cleaned milled white rice was then steamed and coated with glucose and talc in a revolving cylinder to achieve a better finish. Finally, the rice was segregated into head rice and broken rice in large disc graders before packing. The grader unit consisted of a number of cast-iron discs, mounted on a horizontal revolving shaft, the discs dipping into the mass of rice inside a horizontal cylindrical housing. A graphical presentation of the flow diagram in a rice mill functioning during 1940 in Louisiana, USA, is shown in Fig. 5.4. Such rice mills were also exported. Over the period, however, during the next two to three decades, the American design gave way to the European or the Japanese technology (which started becoming popular across the globe), and is practised no more anywhere.

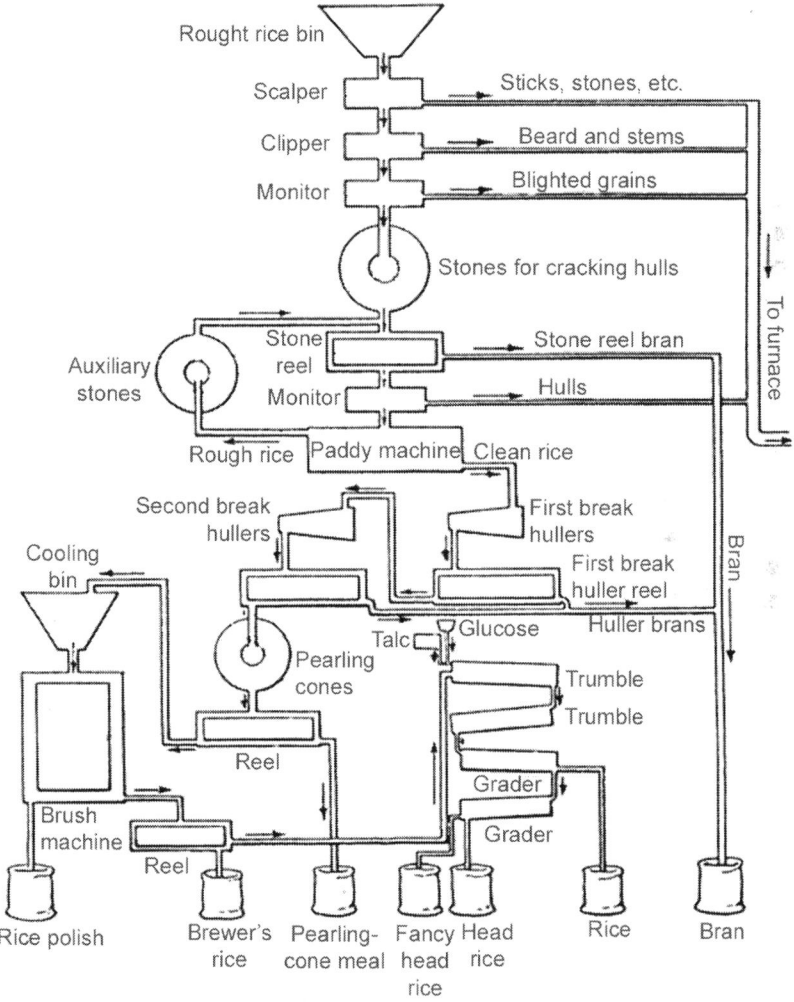

Figure 5.4 Flow diagram of operation in a rice mill in Louisiana, U.S.A., around 1940
[*Source*: Levee 1941]

The rice milling systems developed in Europe were commonly deployed in most of the South Asian countries for the large-scale processing of rice. This was so because majority of the countries were under colonial system of UK or other European countries, till the middle of 20th century. The core machinery comprised of an under-runner disc-sheller, a compartment-type oscillating paddy separator, and a cone polisher. In the paddy separator unit, the separation was achieved based on differences in the surface properties and density between paddy and brown rice. The mixture was made to move in a zigzag path, along an inclined smooth surface. The brown rice stream, issuing from the separator was then passed through an emery-coated 'cone polisher'. Normally, multiple units of cone polisher were employed with decreasing grit size of emery in the coating. A metallic friction polishing unit, like huller, was introduced during 1960s to complete the milling action. Multiple units were also used in series to reduce pressure on the grain in order to reduce the breakage. Grading and separation of rice and broken grains was done either with the use of a set of sieves with different mesh-size opening, or a set of 'trieur' cylinders with different indentation size.

The mill layout spread over a larger area, as the machinery was mostly arranged at ground level. The intake paddy was fed into pits made at a lower level than ground, lifted by bucket elevators and carried to the next machinery by gravity by feeding through pipes. The capacity of the mill was mostly two tonnes per hour. However, one to four tonnes per hour plants also existed. Majority of the mills had a central shaft connected to a high-capacity motor to run the entire mill. Individual units were so arranged that they could be connected to the central shaft by a pulley and belt system as and when required.

Among the traditional rice-producing and -consuming countries, the lead for the development of rice milling machinery was taken by Japan. Many new concepts were, therefore, explored for efficient removal of the husk and bran layers during 1920s in Japan. The most important of them was the development of rubber-roll sheller for dehusking and use of a vitrified, cylindrical emery roll (not the cone), instead of the natural emery-coated cones, as used in European design. The vitrified emery cylinder was housed in a vertical or horizontal cylindrical cage, and had the provision of ventilation in the milling chamber. A combination of high-speed abrasive vitrified rotor, followed by a low-speed rotor, was also developed. This helped greatly in precision milling with very little generation of broken rice. Ventilation system through the rotor shaft was introduced later. This helped in efficient removal of the bran produced in the machine by the milling action. Rubber-roll sheller was developed during 1950s, which brought a revolution in the rice milling technology by elimination of breakage of rice during dehusking. The concept of humidified milling was introduced during 1970s that imparted a shine to the milled grain with complete removal of bran from the grain surface.

Use of rubber-roll sheller for dehusking, and vitrified-emery roll for debranning, improved the quality and yield of brown rice considerably in comparison to the under-runner disc sheller, and cone polisher system. During the same period, Japanese also developed an oscillation-action paddy separator which had a coarse-surface separating tray, oscillating in a reciprocated motion with simultaneous oblique up-and-down action. This resulted in a better efficiency and a higher throughput capacity. A destoning machine was also developed in Japan during 1950s for the removal of sand and stones from the product, which were similar in the size as that of the milled rice.

The Japanese mills, in comparison to the European, were relatively compact and occupied less space for similar capacities of handling and production. The difference in the layout and arrangement of the machinery between the Japanese mills and the other types was noteworthy. Even before the World War II, a European mill was ten times the size of a Japanese mill with the same processing capacity. In the domestic practice, rice milling in Japan was (and still is) carried out in two different phases and levels. Paddy was converted to brown rice, using small mills, by farmers at the farm level itself and the brown rice was collected and processed in big rice mills for the subsequent operations. The land holdings in Japan are of smaller size. Farmers clean, dry, store and dehusk paddy at the farm level itself to produce brown rice. The intake material of large-scale mills was therefore brown rice, not paddy. The unit operation machinery was ergonomically placed in proximity to each other, and the material conveying was done by pneumatic system or through compact bucket elevators. In large mills, a central ventilation system through the shaft, in the debranning units, provided a strong air flow to blow the bran away from rice, and to keep the unit cool.

Grading or separation of broken rice was normally by rotary sifters. Japanese rice varieties are shorter in length, roundish in shape and produce very little broken rice (normally under 5% or so) upon milling. A practice, unique to Japan, is to mix two or more brands or varieties of rice to suit to a desired, particular taste and texture profile when cooked. Therefore, a rice mixing or blending operation was carried out in a 'rice-mixing machine' or the 'rice-blending machine' before bagging.

Japanese mills that were meant for export, however, contained all unit operation machinery streamlined into one complete system from paddy intake to the packaging level. The grit size of vitrified abrasive rolls for debranning was suitably adjusted for handling of 'long-grain' rice.

Development of colour sorter equipment, for the removal of damaged, discoloured and defective grains from the final product, was the next step towards producing high quality milled rice. A colour sorter for removing the discoloured coffee seeds was in fact developed during 1930s in the UK for use in the seed industry and processing of coffee beans. However, a dedicated rice colour sorter, 'Sortex', manufactured in the UK, was first installed in a rice mill to remove the blackened grains in parboiled milled rice, by Uncle Ben's, Inc., in Texas, USA in 1960. An upgraded version of 1965 became more popular and was installed in various rice mills. Further advanced machines are now widely employed in most of the modern rice mills all over the world.

Keeping pace with the innovation and improvement of the main dehusking and debranning machinery, developments also took place in the technology of auxiliary machinery for cleaning, storage, conveying, weighing and packing. Automation, monitoring and control systems were also developed during 1970s and 1980s and incorporated for smooth functioning of the mill, and to maintain a continuous flow of material from intake point to the packaging stage. Recovery of by-products in their pure form and their optimum utilization was also given importance, which improved the economics of the rice milling industry. The normal capacity of large-scale rice mills during 1950s was around 2–6 tonnes per hour. Today, very large rice mills having processing capacities of 100 tonnes to 150 tonnes of paddy per hour are operating in various countries.

5.2   Unit operations and equipment in rice milling

5.2.1   Process flow and unit operations

The purpose of rice milling is essentially to remove the husk and bran from the harvested dried paddy to produce edible and marketable milled polished rice with the least breakage of the grains. The unit operations involved in milling of rice and the basic process flow is shown in Fig. 5.5.

The milling process thus involves the following unit operations which are carried out in sequence.

(a) *Cleaning* to remove the foreign matter and impurities from the paddy stock, such as sand, stones, straw, seeds and pieces of iron, etc., from paddy. This is done normally in two stages. First, a preliminary cleaning is done on arrival of paddy at the mill before storage, and second, at the beginning of the milling operation.

(b) *Dehusking* to remove the husk from paddy with minimum damage to the rice kernel.

(c) *Husk separation* to remove the husk generated during dehusking process, resulting in a mixture of brown rice and unshelled paddy.

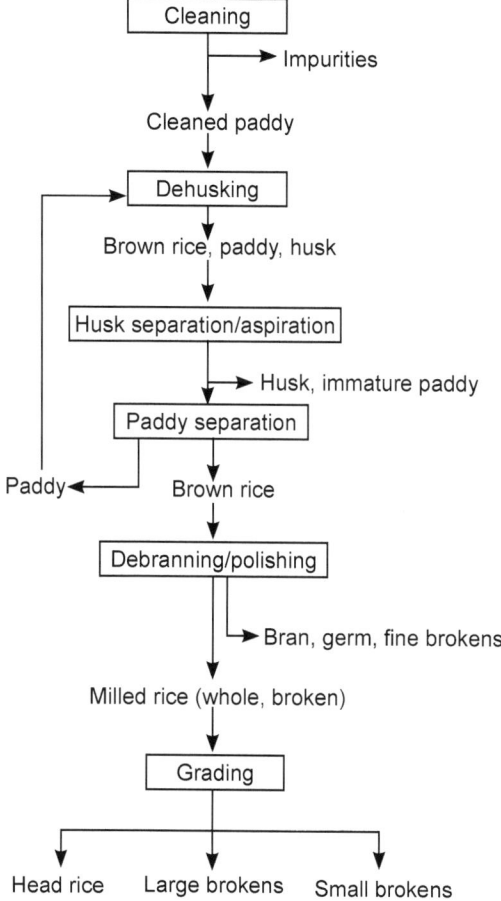

Figure 5.5 Unit operations and process flow diagram for milling of rice

(d) *Paddy separation* which involves segregation of unshelled paddy from the mixture of brown rice and paddy, and return of the separated unshelled paddy to the sheller for dehusking.

(e) *Debranning* and polishing to remove all, or a part, of the bran layers from brown rice surface to produce polished milled rice.

(f) *Grading* to separate the broken grains from whole grains and head rice. The broken grains are further separated into different sizes.

(g) *Bagging and packing* of the finished products into convenient sizes in suitable packaging material for safe storage, transport and marketing.

Each unit operation may require one or more equipment to perform either different functions or the same function in multiple steps. Conveying of material is done by lifting through bucket elevators and delivery by gravity through pipes or chutes to the next unit. Occasionally, screw conveyors or belt conveyers are also employed. Storage bins for the material, at appropriate places, above the equipment are also sometimes used.

The modern, or modernised, rice mills are normally equipped with machinery that have following features:

- Cleaning machines that have provision to remove immature paddy
- Rubber-roll shellers for dehusking of paddy, which results in a higher yield of brown rice, with least damage to the external-most layer of the bran.
- Paddy separators that have controls on all the operational parameters such as speed, angle and oscillation of the deck, as well as feed control, for efficient separation.
- For debranning and polishing of raw rice, the first and second units are of abrasive type (rotor surface made of fine grit silicon carbide or vitrified abrasive rolls), and the third being a friction type, which yields milled rice with a smoother surface. The rotor of the friction-type equipment has a hollow shaft for passage of air through the milling chamber. In case of parboiled rice, abrasive cone / cylindrical polishers with coarser grit size are employed in the beginning step, followed by polishers with finer grit size. Bran is therefore removed in small increments using a series of abrasive mills. However, a friction-type polisher is also frequently used as the last pass for polishing of parboiled rice.
- For grading, plansifter or trieur cylinder is employed for efficient and total separation of broken rice.

A short description of different type of machines used for each unit operation under each stage of milling is given below.

5.2.2   Cleaning

Paddy as received at the mill usually contains some foreign matter, i.e. material other than the paddy grains. This may include parts of straw, soil particles, stones, weed seeds, metal, glass, etc., which come in from the field or during handling and transportation. Cleaning is, therefore, the first step in modern rice milling. It not only enables the production of clean rice but also protects other milling machinery and increases milling efficiency. Differences in the physical properties, like size, shape, density, magnetic conductance, surface and frictional properties, etc., are utilised for separation of impurities and foreign matter from paddy. Initially, impurities based on size, shape and aspiration are removed first, followed by removal of ferrous impurities by magnetic separators. Impurities of the same size of paddy, but heavier, are removed by destoners/specific gravity separators. Some cleaners may have built-in destoners. Majority of these machines are also commonly employed for cleaning of grains other than rice/paddy. The common machines that are employed in the cleaning operation are discussed below.

Scalper

The intake paddy is often subjected to a preliminary partial cleaning (scalping), prior to storage and the main cleaning. Simplest of these machines in which this is carried out is the 'drum sieve'. Paddy is fed into a rotating drum made of wire mesh or a perforated metal sieve. The feed end of the drum is kept slightly elevated. Paddy falls through the sieve openings in the drum, while objects that are much larger than paddy, like parts of straw etc., remain on the sieve and are discharged at the other end. The scalper removes the bulk of large size impurities, but the paddy is not completely cleaned. The small amounts of impurities that still remain are removed later. In some scalp-cleaners, an aspiration system is also added to remove the dust. Figure 5.6 shows the schematic diagram of one such scalp-cleaner.

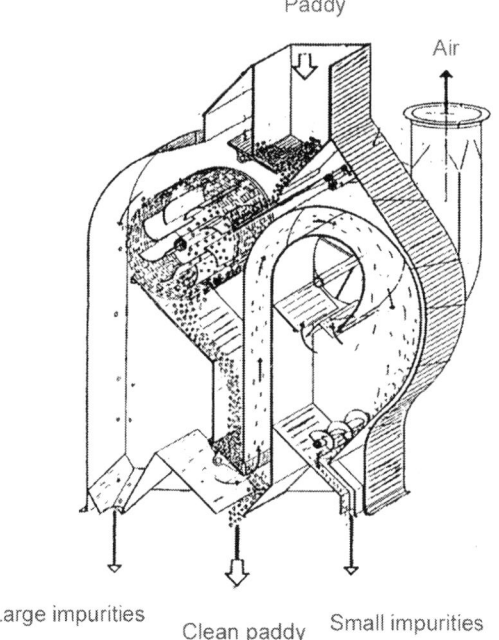

Figure 5.6 Scalp-cleaner [*Source*: Borasio and Gariboldi 1957; reproduced with permission from FAO]

Paddy cleaner

In this machine, Fig. 5.7, paddy is fed through an opening from the top. A suction fan draws air through the film of falling grains and aspirates all dust and light impurities, which are carried to the cone-shaped bottom of the aspiration housing for collection and discharge. The paddy falls onto a series of vibrating decks that are covered with perforated steel sheets. On the top deck, objects that are longer or wider than paddy, such as straw pieces, larger contaminating seeds, stones, etc., are removed. Paddy and impurities of the same size and also shorter than paddy, then fall on to the lower deck which has small perforations. This allows objects shorter than rice, like sand and small seeds, to pass through. This fraction gets discharged separately. Paddy, overflowing from the sieve, is again passed through a strong, second aspiration system, which removes the last traces of light impurities and dust.

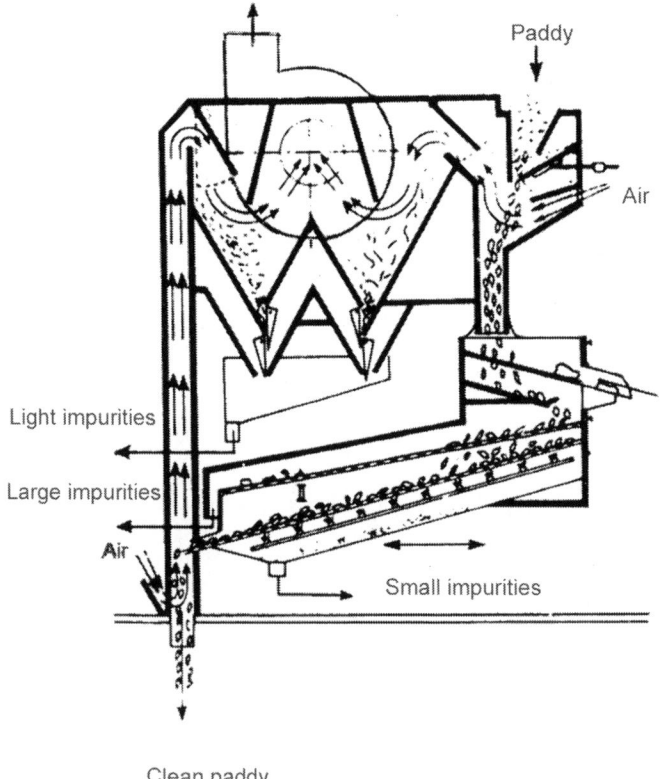

Figure 5.7 Paddy cleaner [*Source*: Araullo et al. 1976; reproduced with permission from IDRC]

Indented disc separators or trieur cylinders (described later) may also be used to separate paddy from longer or shorter impurities, although they are more utilised for size grading and separation of whole grain milled rice from broken grains.

Drum-type cleaner

In this type of cleaner, paddy is first passed through a rotary screen (scalper), which separates the larger particles. The exit paddy from this is then dropped onto a set of vibrating sieves/screen. As paddy is dropping, a fan pulls a stream of air through it removing the chaff and the dust. The vibrating screen/sieves have larger opening on the top sieve and smaller opening in the bottom sieve. This removes foreign material larger than paddy and smaller than paddy. Most often a round-hole top screen and a slotted-hole bottom screen are used. A schematic diagram of such a unit is shown in Fig. 5.8.

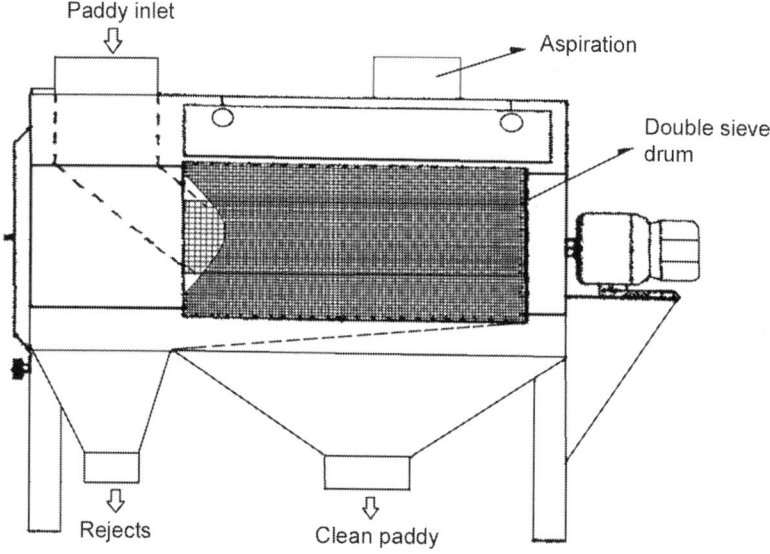

Figure 5.8 Drum-type sieve cleaner

Destoner

After removal of the impurities which are lighter, longer, shorter and wider than the paddy, other impurities such as stones, mud balls, metal or glass pieces about the same size of paddy are removed next by employing a machine called 'destoner' or 'stoner'.

It is a specific gravity separator or density separator. It separates the paddy from stones and other impurities based on the differences in the density. These machines could be of two types, viz. blowing (pressure) type, or suction (vacuum) type.

(a) Pressure-type destoner
This type of destoner (Fig. 5.9) consists of a reciprocating perforated sieve deck mounted at an angle. A large amount of air is blown from below through the sieve. Paddy containing impurities is fed at the top of the sieve. The pressurised air, coming through the sieve with force, stratifies the materials according to their density (or heaviness). The heavier impurities, like stones, which are not lifted by the air, remain on the deck and are carried backward to the higher end of the sieve by the reciprocating motion of the deck and are discharged. The lighter paddy remains floating and slides down to the lower end. The separation can be controlled by adjusting the feed rate of material, air flow and inclination of the sieve.

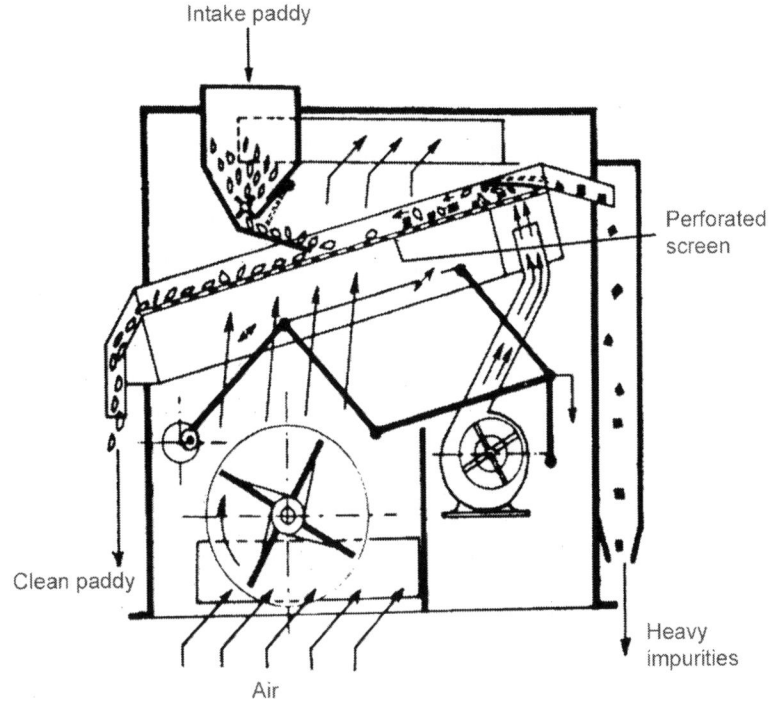

Figure 5.9 Destoner, pressure type
[*Source*: Araullo et al. 1976; reproduced with permission from IDRC]

(b) Suction (vacuum)-type destoner

The working principle in this type of destoner is the same as that of pressure type. The only difference is that the lighter material (paddy) is made to 'float' by the force of suction from the top through an aspirator, instead of being lifted by the pressure of air from the bottom of the deck. Paddy is fed onto an inclined, textured vibrating deck made of woven wire or perforated sheet metal. Air is pulled through the paddy from beneath the deck by powerful aspiration system which causes the bed of paddy to be fluidized (to lift away slightly) from the textured surface of the deck. The vibrating deck pushes the heavier material in contact with the deck upwards towards the stone discharge spout. The lighter material (paddy), which 'floats', flows down the inclined vibrating deck by gravity, and exits through the clean product spout. The inclination of the vibrating deck, the stroke of eccentric motion and air volume can be finely adjusted to achieve an optimum degree of separation. Destoners equipped additionally with an air-recycling system are also available. A schematic diagram of one such unit is shown in Fig. 5.10.

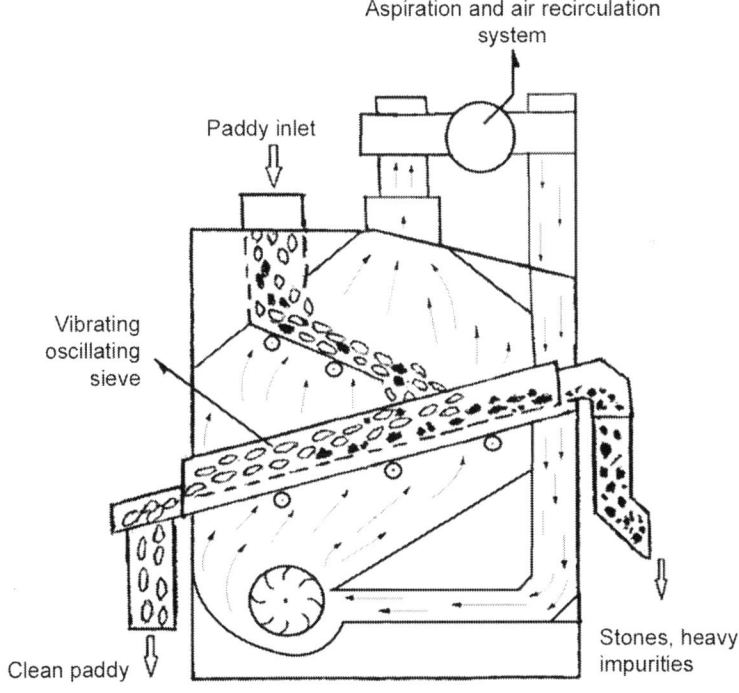

Figure 5.10 Vacuum-type destoner with air recycling system

Magnetic separator

Finally, metal particles, normally made of iron that might have been present in the paddy at the outset, or that could have been introduced into the paddy stream during prior stages of handling and conveying in the mill, are removed by deploying magnetic separators. These are normally of two types. In one type, a permanent magnet is placed in such a way, that when unclean paddy moves across it, the iron particles get collected and remain attached on the surface. The particles are cleared later manually. In the other type, which is normally deployed in high capacity mills, a magnetic drum separator is used. Ceramic magnets are employed in these systems, which are capable of producing a strong magnetic field that could penetrate deep into the bed of paddy. The unit consists of a rotating brass drum or a cylinder (nonmagnetic) which is turned by the free-flowing paddy. Under the cylinder, there is a half-round magnet. As the paddy passes over the rotating cylinder, iron particles are held on the cylinder's surface by the magnetic attraction from underneath. As the cylinder continues to rotate, it moves over to the other half-round where the magnet is not there. The iron particles are then automatically released and discharged separately (Fig. 5.11).

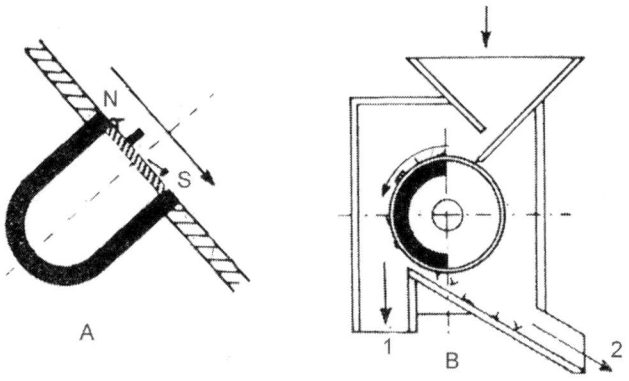

Figure 5.11 Magnetic separator (A – manual; B – continuous)
[*Source*: Araullo et al. 1976; reproduced with permission from IDRC]

Thickness grader

In some modern plants, the cleaned paddy is subjected to grading by thickness in order to remove the immature kernels which are thinner than the normal, mature and fully ripened grains. Optimally harvested paddy may contain normally about 5–7% immature, thin grains. Removal of this fraction renders the main bulk to be of uniform size. This reduces energy consumption during milling, reduces breakage and gives uniform polish and appearance to the milled rice. The thin grains are collected and processed separately with appropriate settings. Removal of immature grains is also important if paddy is meant for parboiling, as immature kernels tend to become dark after the moist-heat treatment during parboiling process. In some plants, however, thickness grading is applied to the brown rice after dehusking step of paddy. The principle of removal of thin grains, however, is the same in both the cases. The grains are tumbled inside a cylinder, or a polygonal cylinder, made of corrugated steel plate, that contains rectangular opening or slots. As the grains continuously tumble inside the cylinder, individual grains eventually line up with the opening of the slot. The thin grains fall through the slot and are discharged from the machine, while the remaining normal and full-size grains remain on top and are conveyed to the end of the cylinder where they are discharged separately. If the slots are made round, and the diameter is suitably adjusted, the machine then serves as 'breadth' grader for the brown rice to separate broader grains.

5.2.3   Dehusking/husking/shelling

After removal of all foreign matter and impurities, the cleaned paddy is ready to be milled further. The next stage of processing is aimed at the removal of husk from the paddy with minimum damage to the bran layer and without breaking the resultant brown rice. The operation of dehusking is carried out in machines termed as 'shellers', 'huskers', or even 'hullers'.

The most common types are the under-runner disc sheller, centrifugal sheller and the rubber-roll sheller.

Disc sheller

The under-runner disc sheller, more often referred to simply as disc sheller, consists of two horizontally placed iron discs, kept at a distance from each other. Their inner surface is coated with an abrasive emery layer. The top disc is fixed to the frame housing, while the bottom one rotates (Fig 5.12). The rotating disc can be moved vertically up or down to adjust the clearance between the two discs. The clearance is fixed depending on the length of the paddy grain. Paddy is fed into the centre of the machine. It moves outward by centrifugal force towards the edge of the disc. It gets evenly distributed over the surface of the disc during rotation. Simultaneously, the grains also get aligned vertically under the centrifugal force and friction of the disc. While doing so, they get caught between the two discs, length-wise, and are dehusked by pressure and shearing action. As the grain has to travel across the entire radius of the disc before it is discharged, some rice breakage does occur in this machine. Some polishing also takes place for the same reason. The clearance between the two discs is critical and requires continuous adjustment to avoid excessive breakage.

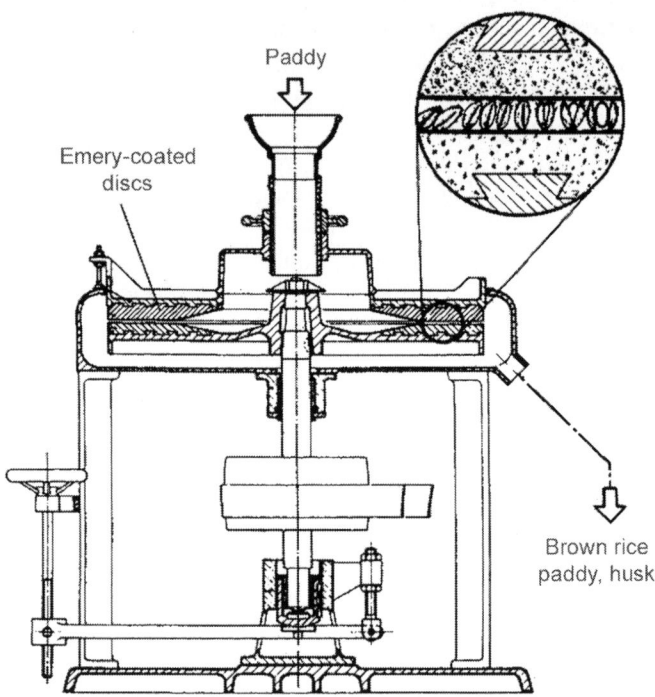

Figure 5.12 Disc sheller (Expanded view shows alignment of paddy grains during the process of dehusking) [*Source*: Borasio and Gariboldi 1957; reproduced with permission from FAO]

The main advantages of the disc sheller are its operational simplicity and low running cost. The abrasive coverings can also be remade at the site with inexpensive materials. The main disadvantages are the grain breakage and a partial abrasion of outer bran layers. This type of shellers were popular prior the introduction of rubber-roll shellers. Their use, however, has now decreased much and they have become almost obsolete.

Centrifugal sheller

This type of sheller (Fig. 5.13) consists of a high speed impeller disc, provided with radial blades, placed vertically and housed in a metal casing with a fixed hard-rubber ring on the inner side. Paddy is fed to the sheller laterally, to the centre of the disc which is rotating at high speed, at about 2500–3000 rpm. Paddy grains, rotated by the blades, move outwards in the radial direction by the centrifugal force and receive frictional force from the blade surface. The grains emerge at high speed like a bullet and the tip collides with the hard rubber ring placed at an angle to the direction of the grains. Dehusking is effected by the impact force.

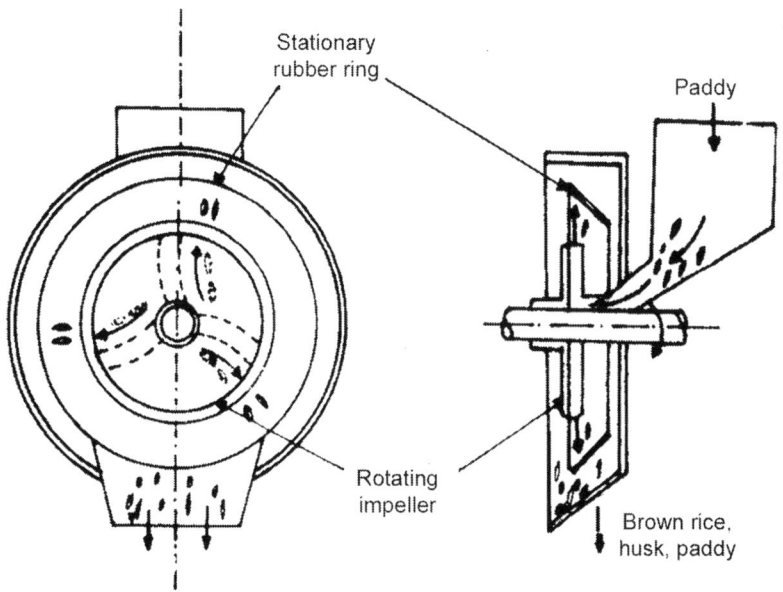

Figure 5.13 Centrifugal sheller [*Source:* Borasio and Gariboldi 1957; reproduced with permission from FAO]

As compared to the disc sheller, performance of centrifugal sheller is better, as it gives a higher shelling efficiency, of up to 98%, and a lower breakage of the grains. However, they are not suitable for handling large capacities and therefore have not found acceptance for large-scale milling. Sometimes, they are used in the small-capacity, improved huller systems, resulting in a better milling performance than using huller alone. Because of their high husking ratio, it is frequently used as a husker attached to coin-operated rice mill units placed at super markets, such as in Japan. The impeller blades and liner in these units are made of plastic material.

Centrifugal shellers do have a distinct advantage over other types, as they perform better for the handling of high-moisture paddy. Even when the paddy contains as much as 24% moisture, the shelling efficiency could be 95% or more. These shellers could therefore be employed for production of certain products requiring handling and dehusking of high-moisture paddy in the process schedule of some special products.

Rubber-roll sheller

All modern rice mills now employ rubber-roll sheller for dehusking of paddy. In fact, it is considered as one of the chief components and a characteristic feature of a modern rice mill.

The unit consists of two rubber-covered rollers rotating at different peripheral velocities in opposite directions. One roller is fixed in position to the housing frame while the other is adjustable and could be slided laterally to increase or decrease the spacing between them. Generally, the distance between them is kept equal to half the thickness of paddy grain. Both rollers have the same diameter, but one roll moves at about 25% higher speed than the other. Due to the difference in the peripheral speeds, the paddy grains falling between the rolls are subjected to shearing and frictional forces that strip the husk off from the brown rice (Fig. 5.14).

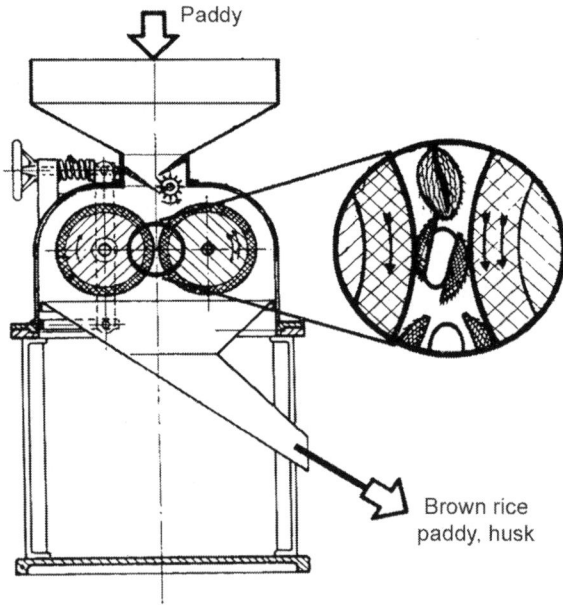

Figure 5.14 Rubber-roll sheller (Expanded view shows the dehusking by shearing action on paddy grain between the rubber rolls) [*Source*: Borasio and Gariboldi 1957; reproduced with permission from FAO]

The rollers are cooled by blowing air on the roll surface. The clearance between the rolls could be adjusted by a hand wheel manually or automatically in the most modern machines, by a pneumatic mechanism with sensors. The unit could also automatically separate the rolls and turn the driving motor off, if the paddy flow through the machine is interrupted. Excessive pressure, by over loading with paddy or keeping the space between rolls lesser than optimal, can cause considerable breakage of the grain. The shelling (dehusking) rate is normally maintained at about 85%. Care has to be taken to see the rolls should not touch each other during operation. Otherwise, there would be excessive heat, excessive wear and tear of the rolls, as well as discolouration of the grains.

The durability or capacity of the rubber rolls varies with cleanliness of paddy, moisture content, and pressure applied to the rolls. It also depends on the paddy variety, and additionally, on the age and quality of the rolls. A pair of good-quality rolls has an average capacity of 100–200 tonne paddy/pair. The capacity is higher with short grains. The optimum age for rubber rolls begins 2–3 months after manufacture and decreases rapidly when the rubber is 6–9 months old. The best performance is shown when the rolls are used between the 3rd and 6th month from the date of their manufacture. Therefore, rubber rolls should be stored only for a limited time. Alternates for natural rubber, to overcome this drawback and to increase the performance, are being explored. White rubber (natural rubber mixed with white silicone) has been developed. Use of synthetic rubber such as polyurethane, which has a high wear-resistance, is also increasing as it imparts durability.

To obtain more operational life per pair of rubber rolls, they should be frequently interchanged to ensure uniform wear. Stationary roll, which rotates at higher speed, wears out faster. It is advisable to interchange the rolls for every 2–3 mm of wear. For optimum performance, single grain layer should be evenly distributed over the full width of the rolls. Otherwise, the roll surface wears out unevenly, reducing efficiency and capacity. Unevenly worn rolls could, however, be corrected by turning them on a lathe.

The new improvements which enhance the life and functioning of the rolls advocate that the two rolls should be arranged obliquely in the unit with respect to each other (Fig. 5.15). The paddy should also be pre-aligned longitudinally and distributed evenly on the roll surface while it is entering between the rolls. It is claimed that this arrangement gives a higher shelling efficiency and lower wear of the rolls, ensuring a longer rubber-roll life.

Compared with the other type of shellers, the rubber-roll sheller has the advantage of causing negligible breakage and no damage to the bran layers even at 85% shelling rate. In the disc sheller, on the other hand, shelling beyond 60–70% causes considerable breakage, and some fine broken rice may even be blown off along with the husk. However, the rubber rolls wear out fast and have to be replaced often. This undoubtedly is a disadvantage, but is offset by the reduction in breakage and increase in the total and head rice yield.

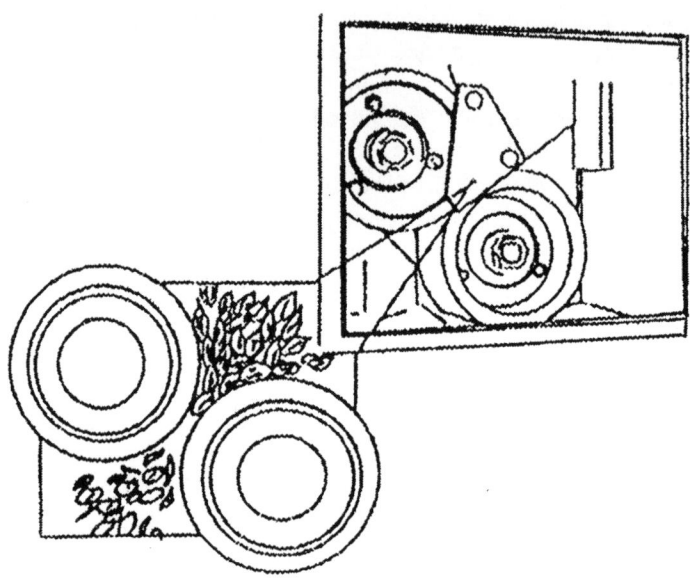

Figure 5.15 Oblique arrangement of rolls in the rubber-roll sheller
[*Source*: Ishii 1995]

5.2.4   Husk separation

Depending upon the type of sheller used for dehusking of paddy, the material streaming out of the unit would contain different fractions. In the case of disc sheller, the output consists of a mixture of brown rice, remaining unshelled paddy, some broken rice, husk particles, and sometimes even bran and germ. These fractions are separated based on size (by appropriate sieves), density and frictional properties (by aspiration, movement on a surface).

The broken grains, germs, and bran can be separated through an oscillating sieve with fine perforations, and discharged separately. Husk, being lighter than any of the other particles, is easily separated with aspiration. The two separations can be done in one machine by the provision of an oscillating sieve and an air aspirator. A plansifter equipped with two self-cleaning sieves, one with fine perforations for separation of bran and dust, and another with larger perforations for collection of broken grains is normally used (Fig. 5.16). The over-flow from the sieves contains husk, brown rice and unshelled paddy. In the lower part of the machine, a husk aspirator pulls air through the mixture and lifts the husks to be discharged through the blower opening into a cyclone. The immature grains are also lifted, but get dropped at a certain height to a separate collection chamber. The air could be forced through the falling material from below (blowing type) or drawn through the top (suction type) in these aspirators. A mixture of brown rice and paddy flows out of the unit to a paddy separator.

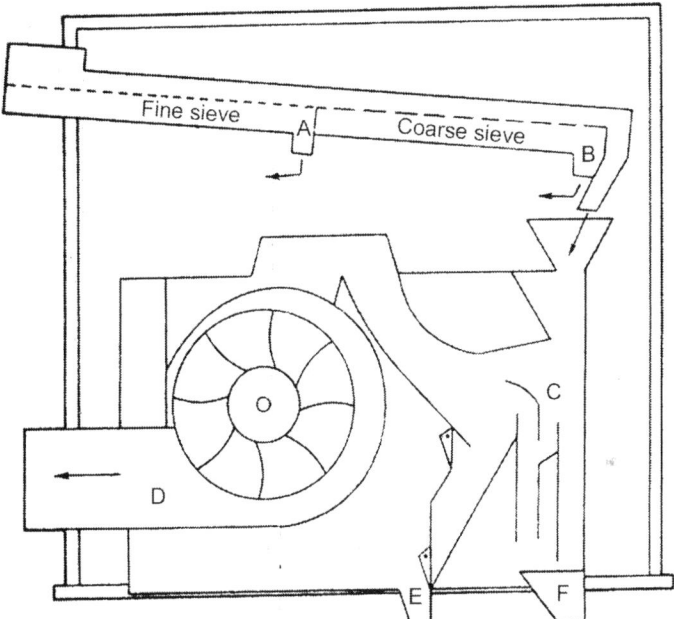

Figure 5.16 Plansifter with husk aspirator (A – bran and dust; B – broken rice; C – air mixing chamber; D – husk; E – paddy; F – brown rice) [*Source*: Wimberly 1983; reproduced with permission from IRRI]

In case of rubber-roll sheller, there is no damage to the bran layers and the discharge does not contain any bran, but only a small content of broken grains. It is therefore directly connected to the husk aspirator unit. Therefore, following the rubber-roll sheller, a plansifter is not required, and only a husk aspirator is used. Schematic drawing of a typical husk aspirator with a rubber-roll husker is shown in Fig. 5.17.

Figure 5.17 Rubber-roll sheller with husk aspirator (A – paddy; B – brown rice and paddy; C – immature paddy; D – husk) [*Source:* Wimberly 1983; reproduced with permission from IRRI]

Another type of husk aspirator often used with rubber-roll sheller is the 'closed-circuit'-type separator. It is called as a closed-circuit separator because it does not blow out the husk and air. The air is again used for separation and is continuously re-circulated. In these units, air is pulled through the stream of falling output layer, lifting out the husk, which settles in an expansion chamber and is discharged by a screw conveyor. Paddy and brown rice also get discharged by a separate screw conveyor, while the air is continuously re-circulated.

5.2.5.   Paddy separation

About 10–15% of the throughput coming out of the husk separator is unshelled paddy that comes out mixed with brown rice. Paddy needs to be removed before the brown rice goes for bran-removal stage. These two constituents of the mixture are separated using 'paddy separator'.

The separation is based on the differences between their certain physical characteristics. In comparison to brown rice, the density of paddy is higher, and the paddy grains are longer, wider and thicker. The surface of paddy is rough, while that of brown rice is smooth, i.e. their coefficient of friction is different. These differences are taken advantage of in separating the two grains. There are two types of paddy separators that are commonly used for the purpose, the compartment type and the tray type.

Compartment-type separator

The compartment-type separator (Fig. 5.18) is the older machine, patented in Europe during 1892, and has been in use for more than a century in the grain-processing industry. The main part is the oscillating compartment assembly in which the separation takes place. The assembly is made of wood or steel, with one or several stacked decks containing many compartments. The number of compartments and the decks determine the capacity of the unit. Each compartment is divided into a zigzag channel and is inclined from one side to the other along the length of zigzag channel.

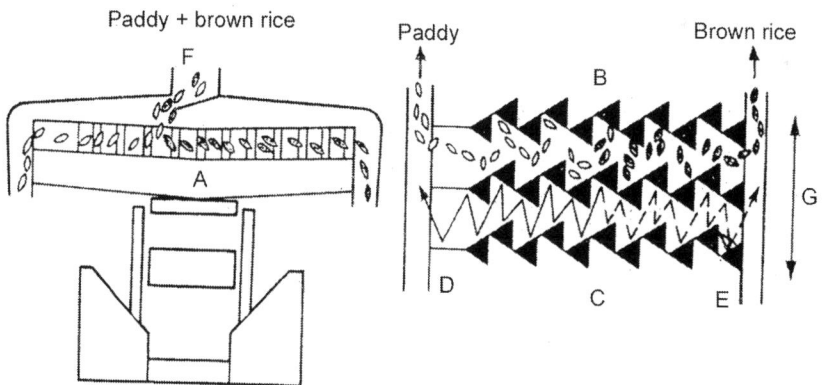

Figure 5.18 Paddy separator – compartment type, cross-section and tray details (A – oscillating table; B and C – zig-zag compartments; D – high side of the table; E – low side of the table; F – paddy input hopper; G – oscillating directions) [*Source*: Wimberly 1983; reproduced with permission from IRRI]

The surface of the table, i.e. the bottom of the compartment, is made of smooth steel. The table is made to oscillate cross-wise, i.e. perpendicular to the direction of the grain flow. The mixture of brown rice and paddy is fed from the hopper to the centre of the channels. As the table oscillates, the grains are thrown from side to side across the channel. The impact of the grains on the sides of each channel causes the unhusked paddy grains to move up the inclined slope towards the higher side of the table. The dehusked brown rice, having higher density, slides down the slope over the smooth surface to the lower side of the table. The output of the separator therefore consists of two streams, one of pure paddy and the other of pure brown rice. The slope and the stroke of oscillation are adjusted to meet the needs of paddy of different size and shape, to ensure complete separation.

This type of separator has the advantage of low power consumption, low operating cost, and low maintenance cost. The tray bottoms and the compartment zigzags can be replaced locally as and when they wear out. The disadvantages of the machine, however, are that it is bulky, requires a strong foundation, and takes considerable space in the mill.

Tray-type separator

This type of separator is becoming more popular in the recent times. It consists of several indented trays (i.e., trays having a large number of depressions) mounted one above the other, about 5 cm apart, all attached to an oscillating frame (Fig. 5.19). The trays are inclined both to the front (long axis, about 4–9°) and to the side (short axis, about 6–11°). When the trays oscillate, the mixture of brown rice and paddy, which is fed at the higher end of the tray, spreads to form a thin layer. There is a slight jumping movement imparted to the tray as it is made to move up and forward. Brown rice, having smaller size but a higher density than paddy and with a smooth surface, moves to the lower level due to the percolation effect. As it contacts the tray surface, it gets caught in the indents on it and is conveyed upward by the upward motion of the oscillation stroke. It gets concentrated on the distal, higher end of the tray from where it is discharged and sent to the polisher. On the other hand, paddy 'rides' and 'floats' on the top of the brown rice layer due to the repelling effect of the smooth surface of brown rice. It slides downwards and accumulates at the lower end of the tray, from where it is discharged and conveyed back to the sheller for dehusking as 'return' paddy. A mixture of paddy and brown rice covers the middle part of the tray, which gets discharged from the tray in the in-between portion, and is returned to the hopper of the separator for recirculation.

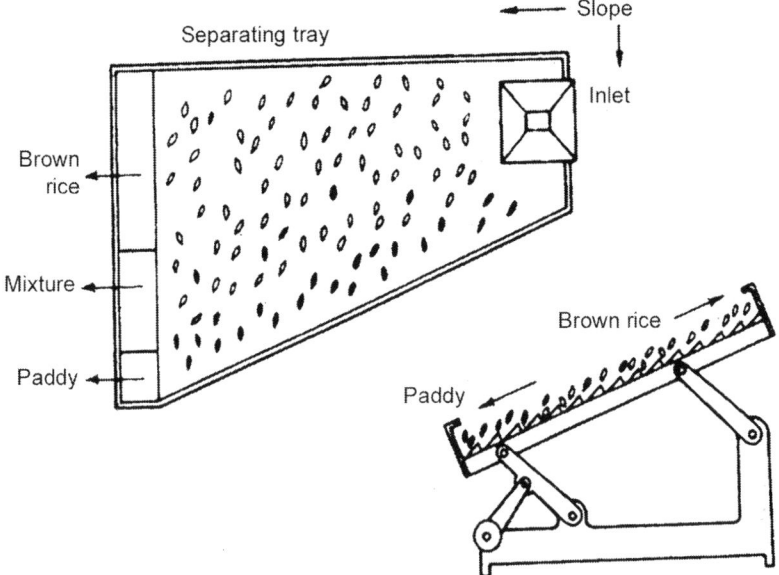

Figure 5.19 Paddy separator – tray type [*Source*: Wimberly 1983; reproduced with permission from IRRI]

The inclination of the deck is adjustable to meet different grain sizes and conditions. For efficient paddy separation, feed rate, speed, stroke and table inclination should be carefully set. Improper settings lead to brown rice going with return paddy or return paddy going with brown rice, or both, affecting plant capacity, the yield and quantity of milled rice.

The all-steel construction, low power requirements, and simple operation lead to low operating and low maintenance costs. The indented steel plates require replacement after long years of use. Another advantage of the tray separator is its small space requirement. This makes the mill more compact and saves floor space.

5.2.6   Shelling of return paddy

Paddy collected from the separator is called the return paddy, for it has to be returned to the sheller. The grains in the return paddy are shorter (if it is from disc sheller) or thinner (if it is from rubber-roll sheller) than the bulk of the paddy. Hence it is preferable to collect the return paddy separately and shell it at the end with appropriate settings. Alternatively, the return paddy could be shelled in a separate, small sheller, adjusted to suit its thickness. This would increase the efficiency of the plant. If the return paddy is returned to the original sheller, with the same settings, it will simply go on circulating, thus lowering the capacity of the plant.

5.2.7   Debranning/whitening/polishing

The brown rice, free from paddy, is then subjected to the debranning in the next stage, in which the outer and sometimes also the inner bran layers are removed. The process of debranning the dehusked rice is referred to by different terminologies in different parts of the world, adding confusion to the meaning and understanding of the process. It is termed as 'whitening' in the American and European descriptions, possibly due to the fact that removal of the coloured bran layers yields the inner 'white' endosperm. The term 'polishing', in such cases, refers to another specific process, which used to be given at the end of 'whitening', wherein the milled rice was passed through a rotating drum fixed with leather straps on its surface. The process also included addition of talcum powder / calcium carbonate and glucose to give the rice a 'fine' finish with a glazing surface. However, after the introduction of the new 'water jet friction polisher' in the recent past, this practice has almost disappeared. In Japan, although the term 'whitening' or 'polishing, is used for the abrasive debranning step, the subsequent friction polishing is termed as 'pearling'. The term 'polishing' there refers to a subsequent process of removing small bran particles, that stick to the rice surface, by water jet polishers that gives the milled rice a glaze and shiny appearance.

In the Indian subcontinent, the term 'polishing' applies to debranning by both abrasion as well as friction process, including the application of humidified friction milling.

The term 'bran' is in fact a collective term applied to the outer layers of brown rice consisting of pericarp, seed coat, aleurone and sub-aleurone layer and also the germ. Some amount of bran removal is necessary for easy cooking and digestion. Excessive removal, however, reduces the nutritional value of rice.

Two types of processes, viz. abrasion and friction (which are associated with the names denoting the respective equipment), are normally employed for the removal of the bran from the grain surface, as illustrated in Fig. 5.20. In the case of abrasion, the grain surface is cut and abraded by emery, and in the second, the grains are subjected to pressure under motion, the friction causing peeling off of the bran layers. The abrasive roller acts like a blade that cuts and removes small bits of bran layer from the brown rice. The process is similar to cutting off an orange skin, little by little in small pieces, with a sharp razor blade. In friction type of machines, on the other hand, bran is removed as big flakes, similar to peeling of the skin of an orange.

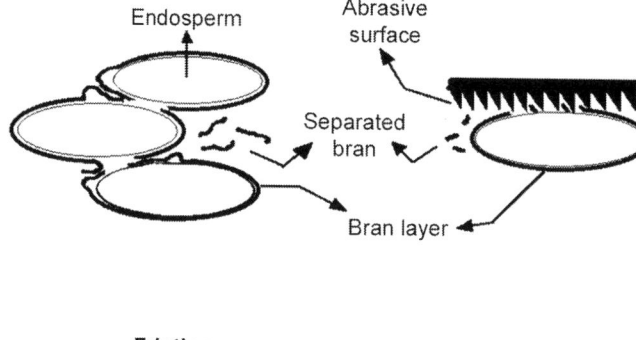

Friction   **Abrasion**

Figure 5.20 Principles of debranning

It may, however, be pointed out that friction, cutting, grinding, pressure and impact forces may all be in operation together in different types of machines, but they are denoted and characterised by the predominant ones.

The surface of the milled rice emerging from the abrasive type of polisher is rather rough, and not very smooth. In case of friction type, since the peeling is done by compression and rubbing, the milled rice has a smooth surface, which has a better market and consumer appeal. In the modern rice mills, therefore, it is an accepted practice to employ abrasive polishers for the initial stages of debranning, followed by frictional polishers.

The abrasive polishers are also of two types – vertical and horizontal. And in vertical type there is the traditional 'cone polisher' and the newer 'cylindrical' polisher.

Abrasive polishers

(a) Vertical 'cone' polisher
It consists of a vertical, truncated inverted metal cone, covered with emery to give it an abrasive surface. The cone rotates inside a wire-mesh screen or perforated sheet housing. The clearance between the cone and screen can be adjusted by raising or lowering the cone. At regular intervals around the cone, the wire screen housing is divided into vertical segments between which rubber brakes are placed at regular intervals. The brakes extend the full length of the cone and project into the space between the cone and the screen, but clear the cone surface by about 2–3 mm. The sides of the machine are provided with doors for periodic inspection of the screens and adjustment of the breaks by means of a hand wheel. Air is drawn from the milling chamber housing, through an opening from the top, to remove the moist warm air generated by

the heating of rice in the milling chamber. The peripheral speed at the middle of the cone is normally adjusted to around 13 m/s.

The brown rice is fed at the centre of the cone from the top. It moves outward by the centrifugal force to the cone, and then moves downwards, rotating into the space between the cone and the screen. Due to the swirling motion, enhanced by obstruction to their flow by the rubber brakes, the grains are brought in contact with the rotating emery surface and the bran gets cut and abraded off the grain surface. Certain amount of friction is also created due to rubbing action between grains, as well as with the surface of the screen. The separated bran goes out through the screen and is collected separately. The polished rice is discharged at the bottom (Fig. 5.21).

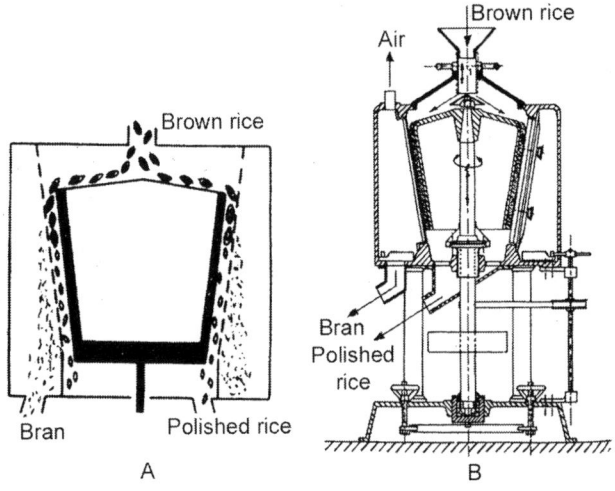

Figure 5.21 Cone polisher (A – principle; B – sectional view) [*Source*: A – Wimberly 1983, reproduced with permission from IRRI and B – Araullo et al. 1976; reproduced with permission from IDRC]

The cone diameter, height, rpm and the rubber-break depth in the milling chamber are all adjusted such that the resultant centrifugal and gravitational forces cause the grain to spiral down for scouring of the rice all around. Removing all the bran in one step causes much breakage to the rice and reduces total rice recovery. To minimize these drawbacks, more number of cones in series is employed in most of the modern mills.

In recent development, vitrified cones are also being manufactured using silicon carbide mixtures. The cones come in sections of 2–4 pieces, depending upon the total height, and are claimed to give better performance than the conventional cones.

(b) Vertical 'cylindrical' polisher

The vertical whitening machine is in fact the improvisation of the European design of cone polisher described above. It is now made by many manufacturers all over the world and is being used increasingly in modern rice mills. In this type of polisher, the brown rice is introduced in the annular space between an outer perforated metal cylinder and an inner rotor cylinder, composed of abrasive wheels packed on a central shaft. Depending upon the direction of rice flow in the unit, vertical polishers are also of two categories. In one, which is more popular, rice grains are fed from the top, move downwards by gravity and emerge at the bottom after they undergo abrasive action in the annular space (Fig. 5.22). In the other type, rice grains are introduced in the milling chamber at the bottom and are pushed upwards as they undergo abrasion, and get discharged from the top port.

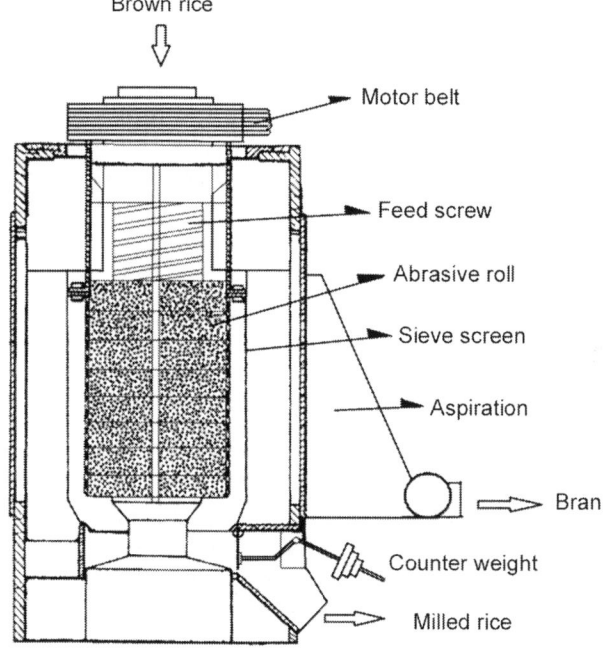

Figure 5.22 Vertical cylindrical abrasive polisher [*Source*: Buhler 2014]

In general, vertical mills receive rice grains under uniform pressure. In the machine that pushes rice grains upward, the grains are loaded at the bottom of the unit by a horizontal conveying screw and then are pushed up with the help of the screw roll provided in the abrasive section. However, the mechanism makes it difficult to control the pressure in the milling chamber and also to discharge rice uniformly from the circumference of the milling chamber at the top. In the machine that uses downward flow of the grains, brown rice enters the milling chamber by a feed screw and moves down by gravity, receiving abrasive action by the rotating cylindrical emery stone. The pressure (and hence the degree of bran removal) is controlled by hanging different weights on the discharge gate. Air is sucked through the milling rotor and the chamber, which prevents development of heat, reduces breakage and removes the bran being produced continuously.

(c) Horizontal abrasive polisher

The horizontal-type abrasive polisher was developed in Japan after World War II. It is well suited for the short-grain Japanese varieties of rice. However, its introduction in other countries of Asia has been slow, where long and medium grain varieties are prevalent, because sensitive adjustments in the machine are necessary to process such varieties. Compared to the sturdy, solidly constructed vertical cone polisher, the horizontal abrasive polisher is more compact in its construction details. Basically the machine (Fig. 5.23) consists of a cylindrical abrasive roll, formed by packing of abrasive discs, clamped on to a horizontal shaft. The cylinder runs at high speed, in a cylindrical chamber made of slotted perforated sheet. A feeding screw fixed on the horizontal shaft feeds the rice into the free space between the abrasive roll and the slotted cylinder. A counter pressure is built up in the milling chamber by a gate valve on the discharge outlet to which weights are added.

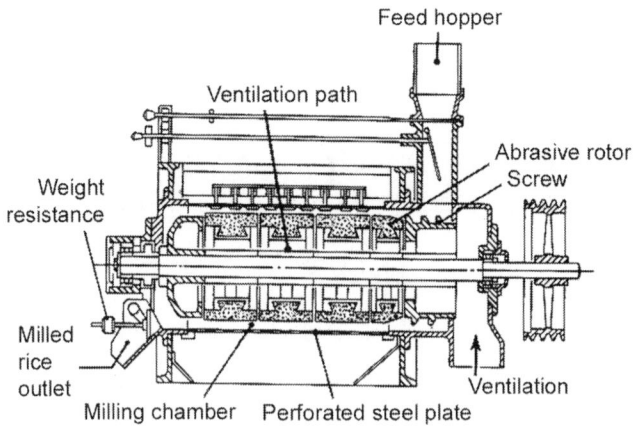

Figure 5.23 Horizontal abrasive polisher. Reproduced, with permission, from Satake Group, Japan

Over the full length of the cylinder, three or four rows of adjustable steel brakes are installed that can be adjusted from 0° (axial) for short and rounder varieties, to 90° (radial), for medium- and long-grain varieties. These adjustable brakes direct the position of the rice grain inside the machine during the scouring process to obtain optimum debranning efficiency. For a long-grain variety, at 0° position of the brakes, the grains are lined up in an axial direction and roll under pressure through the milling chamber. For short-grain varieties, at the 90° position of the brakes, the rice is lined up in a radial direction and tumbles through the pressure chamber of the machine. If long-grain rice is processed in the machine at the 90° position of the brakes, the grain will be badly broken.

An air stream is blown through the hollow shaft and through the small openings provided between the abrasive emery discs. The air passes through the rice and the perforated screen. This keeps the rice temperature lower, thus reducing breakage and also helps to remove the bran sticking to the grains or to the machine.

A disadvantage of this machine, however, is that the clearance cannot be adjusted. Once the diameter of the 250-mm roll disc is reduced by about 6 mm or so, it must be replaced. The rolls, which are made of carborandom/silicium carbide grits of 32 mesh size, cannot be recoated, and new supplies of the discs from the manufacturers have to be secured continuously.

Friction polishers

(a) Horizontal friction polisher

This unit is also called a jet pearler or pneumatic pearler. The unit consists of a cylindrical steel roller rotating inside a hexagonal perforated screen housing (Fig. 5.24). The cylinder has long slits along its length for passage of air. The rotor has ridges protruding from the surface along its length, forming 90° angle at its base on one side, and a solid curved slope on the other. The shaft is hollow for passing of air. A short worm conveyer propels the rice, fed through the hopper, under slight pressure into the unit. The discharge port at the other end is covered by a hanging plate on which weights are added to offer resistance to discharge and to develop back pressure for effective peeling of bran layer. The clearance between the screen and cylinder is adjustable by opening or closing the screen. A strong stream of air is blown by a centrifugal blower through the hollow shaft and the long slit of the cylinder, through the rice and out through the screen. The air helps in separation of bran from rice and also in removal of the heat generated by the friction between rice and rice.

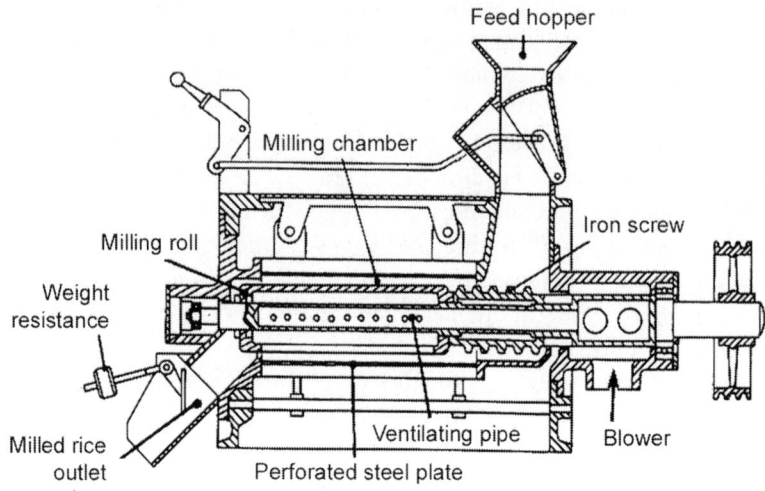

Figure 5.24 Horizontal friction polisher. Reproduced, with permission, from Satake Group, Japan

Steel hullers are also used some times as friction-type polishers, particularly for parboiled paddy. Generally, abrasive polishers are used as the primary polishers and the friction polishers as the final ones. The percentage of bran removed by abrasive polishers is the highest. Friction polishers remove the adhering bran and impart a glaze to the rice surface.

Removal of bran in stages by passing rice through several polishers in succession, gives minimum breakage during milling, thus increasing total and head rice recovery. Keeping uniformity and proper balancing of abrasive roller also reduces grain breakage. When parboiled rice is milled, the bran tends to stick to the screens. In such cases, the quantity of aspirated air is increased to overcome clogging of the screens. The bran coming out of the polisher should be checked often to make sure that it does not contain broken rice or head rice and to ensure that there is no damage to the screens.

Recently, systems have also been developed which consist of multiple abrasive machines and a friction machine packed on below the other. In this case, with multiple units connected, the rice flows directly to the next machine without using an elevator. It also has the advantage of the system being compact.

(b) Vertical friction polisher

This type of polisher consists of a vertically positioned cylindrical steel roller rotating inside a perforated cylindrical screen housing (Fig. 5.25). Rice is fed into the milling chamber at the lower end by a screw feeder and is pushed upward in the cylindrical chamber between the rotor and the screen. Pressure inside the milling chamber is created by putting weights on the outlet hanging gate at the top end. Rice passes from bottom to the top of the chamber and is milled by friction. High pressure air is blown through the chamber that removes the bran generated during milling and keeps the temperature of the chamber low, thus preventing higher breakage of rice.

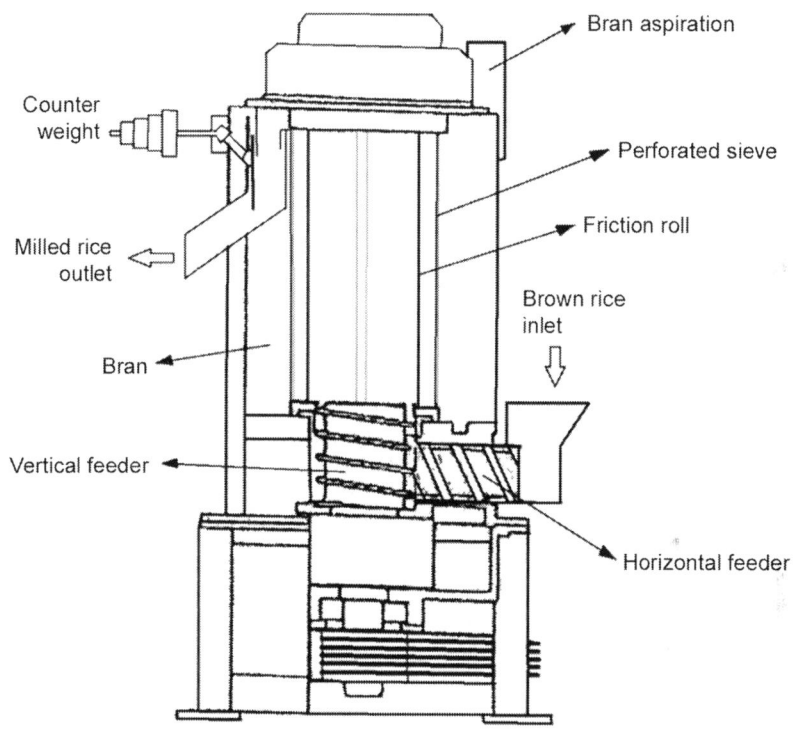

Figure 5.25 Vertical friction polisher

(c) Combined abrasive and friction polisher

Recently, another type of rice mill has been introduced in which both the abrasive and the friction polishers, in sequence, are combined in one unit. In one of the design (Fig. 5.26), rice is first fed through a screw conveyer to the bottom of the vertical abrasive section. The grains are pushed upwards through the abrasive section and emerge from the top. Rice is then conveyed to the bottom of a second vertical friction section in which rice grains are again pushed upwards while undergoing friction between a metallic roll and perforated screen. The pressure is controlled by counter weights at the discharge outlet.

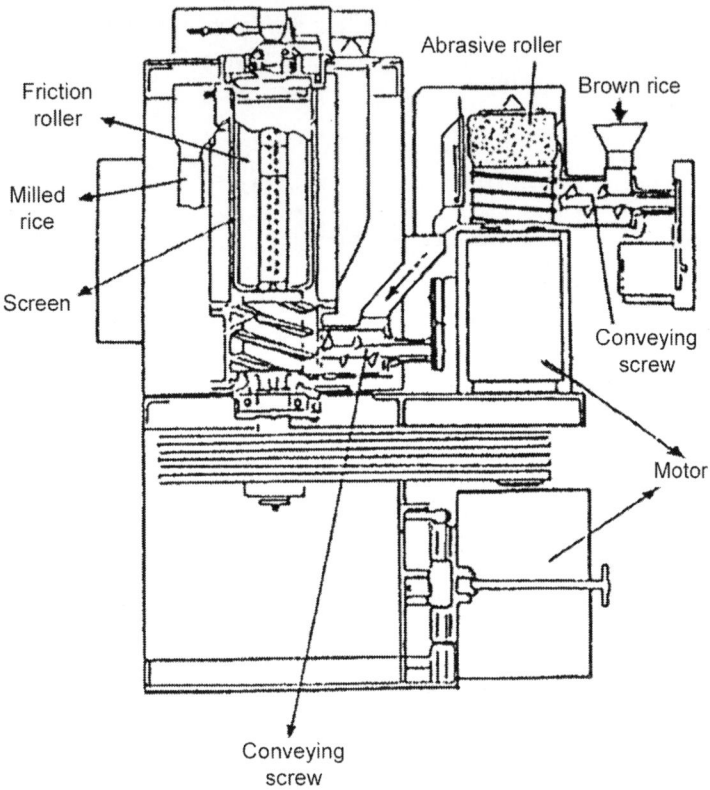

Figure 5.26 Combined abrasive and friction mill, grain movement is upwards in both the sections [*Source*: Katsuragi 1995]

Of late, there is another milling system that has been introduced in Japan, in which rice is passed through three separate milling sections. In the first, a vertical abrasive cylinder polisher, the rice is passed from top to bottom of the unit. In the second, a vertical friction cylinder, the rice is made to move upward. The third one is the same type as the second one except that water-moistening equipment is added to it. It is claimed that this arrangement gives lower breakage and a slightly higher yield than when the water moistening is avoided.

In a yet another design, developed in Japan, both abrasive and friction polishing section are on a single vertical rotor as one compact rice milling unit (Fig. 5.27). The grains are fed uniformly from the top and move downwards through the upper dough-nut-shaped abrasive section first. The rotating force of the grind stone and the bran-eliminating screen are said to even out the rough surface of the grains. The rice then passes to the lower friction section, which has a metallic roll in perforated screen housing with a pressure control mechanism. With appropriate settings, there is little generation of heat, breakage is low and the quality of rice is said to be good.

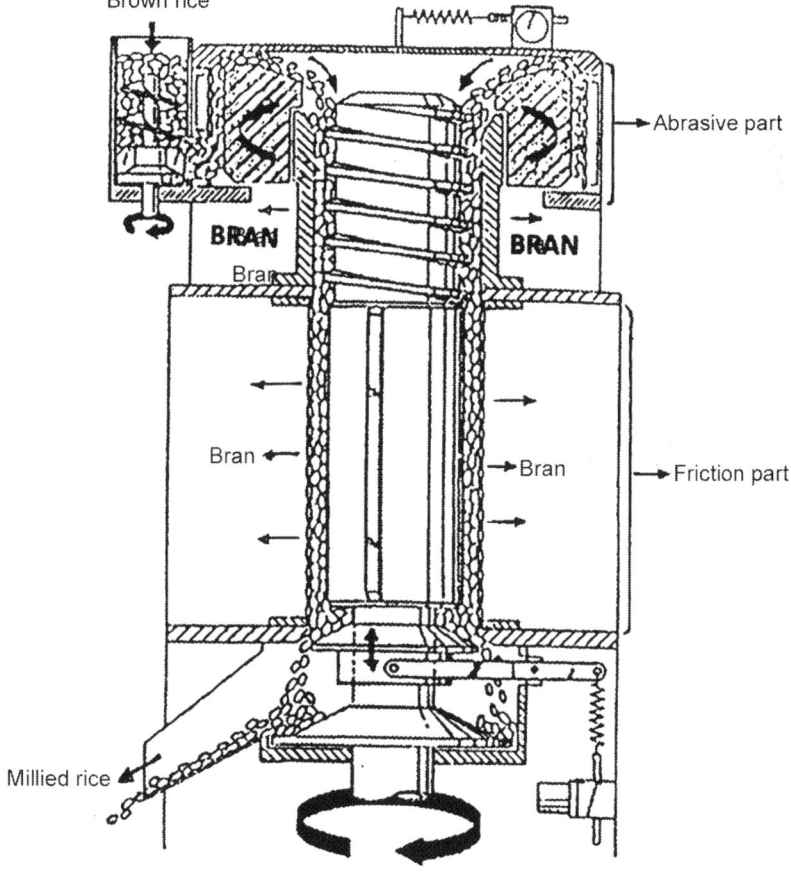

Figure 5.27 Combined vertical abrasive and friction polisher on the same rotor
[*Source*: Ishii 1995]

Water jet polisher

The unit is also called as 'humidifying rice milling machine' or 'humidified milling polisher' (Fig. 5.28). It is used at the end of debranning process for producing milled rice that is fully polished and has a very smooth, shiny and glossy appearance. The rice is sometimes termed as 'silky' rice. A fine water mist is applied under pressure in this machine, prior to polishing. Basically, the unit is very similar to the horizontal friction polisher described above. Rice, that is almost completely milled, is fed from the feeding hopper. A spiral rotor conveys the rice to the milling chamber. A strong air flow from the main hollow shaft, and a water mist from the nozzle enter the milling chamber. Rice is made to move quickly in a circular motion around its feeding axle, grains rubbing against each other under high pressure and very high speed. The brief humidification softens the surface of the grain. The action of air pressure and the friction force not only removes the dust and bran, but also the aleurone layer remaining in the bottom of the longitudinal grooves on the rice grain surface. The amount of water added to the rice during humidification, however, has to be precisely controlled. Some systems provide additional moisture-conditioning units, before the water jet polisher, for use with specific varieties depending on the thickness of bran layers. Although humidification during milling process was long disregarded for fear of cracking the rice grain, safe humidification finally was achieved through the proper combination of speed, pressure and air force.

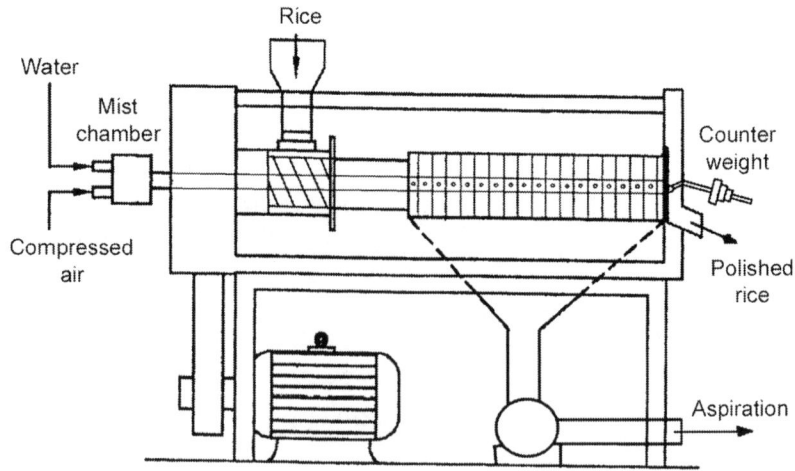

Figure 5.28 Water jet polisher

Use of this unit is becoming more popular as it gives a product that has a better consumer appeal and storage properties.

In a further recent development in Japan, a new fine ceramic material has been used for polishing unit, which does not have the drawbacks associated either with abrasive polisher or the friction polisher. The friction polisher tends to generate heat and broken rice, while the abrasive mill is claimed to damage the surface of the rice grain leading to more adhesiveness in cooked rice. On the other hand, the ceramic rotor has very fine blades on its surface that 'scrap-off' the bran layer. Only the skin is peeled off and there is no grinding or abrasive action. This is achieved effectively at low pressure and low temperature. The starch portion is neither damaged nor scratched. The unit is further combined with a newly developed bran-moistening system. Thus, the new system has the arrangement of a ceramic 'scrape-off' polisher at the top, a frictional polisher in the middle and a bran-moistening system at the bottom of the polishing line. Further, the rotor of the friction polisher is modified and contains boomerang-shaped ridges on its surface. This makes the grain not to slip, and changes the pressure distribution of grains in the milling chamber. The bran is eliminated by centrifugal force. It is claimed that water jet polishing, described earlier, changes the texture of the rice surface as water is applied directly on the rice grains. This results in an inferior taste of the rice when cooked (in Japanese context). In the new bran-moistening system (moisture conditioning of the bran at low temperatures), moistened bran balls are used to absorb fine powdered rice bran. The action is similar to the cleaning of wooden floor by spreading damp tea leaves on it to absorb the fine dust. The rice is thus effectively polished at low temperatures, until there is no trace of bran. It is claimed that the taste of cooked rice does not deteriorate, as the grain surface is not damaged during the new system of milling.

5.2.8   Bran separation

The bran generated during milling from the debranning and polishing units is collected and processed to separate fine broken rice and germ from it. In traditional rice mills, the collected bran is passed through reciprocating sieves for separation of theses fractions. In the modern mills, both the abrasive and the friction polishers are equipped with air blowing or aspiration system. The bran, therefore, gets continuously discharged from the units and is led to the pneumatic bran separation system with cyclone collectors. A blower and an auxiliary cyclone at end aspirate the bran from the main cyclone, thus separating bran from germ and fine broken rice.

5.2.9 Grading

The milled rice coming out of the last milling operation contains not only the 'whole' grains, which form the bulk of the output, but also broken grains of different sizes. Sometimes, if the last milling machine was not the water-jet polisher, it may also contain some bran and dust. In such cases, bran and dust are removed first before grading. For this, the rice is passed through an aspiration system fitted with cyclones. A blower with an auxiliary cyclone aspirates the bran from the outlet of the main cyclone. The fine bran and the coarse material (consisting of germ and very fine broken rice, if any) are thus collected separately.

Broken rice is segregated into different fractions according to its length. Fraction consisting of unbroken whole grains, along with broken grains that are three-fourth or longer in length than that of the whole grain, is classified as 'head rice'. Broken grains that are one-eighth or less in length than that of the whole kernel are classified as 'fines' or 'fine broken'. The rest in between are further classified into 'large' and 'medium' broken depending on their length.

Broken rice is separated by using either oscillating sieves or a plansifter having a deck of sieves, or using an indented cylinder (trieur) or disc separator.

Oscillating sieve

The unit consists of a horizontal screen made of perforated steel sheet with round holes, or wire mesh, set in a frame at an angle of about 4° to 12°, and is vibrated to and fro by a vertical eccentric drive. Single or double screens are used with different sieve openings along the length of the unit. The single screen separates milled rice into head rice and the large broken rice on the upper section, and small broken rice at the lower. The double screen separates milled rice into head rice, large broken rice and small broken rice through the individual screens/sieves. The limitation of this system is that the capacity of the unit is not large, and the effectiveness of separation is also not so high. The machine, however, is widely used in small rice mills in Southeast Asian countries.

Plansifter

A plansifter is generally a screen tray suspended by four steel cables (Fig. 5.29). It consists of a single or double deck sieve which is given a swinging motion produced by an eccentric drive. The sieves, or perforated sheets, contain opening of different sizes to separate milled rice into head rice, large broken rice and small broken rice, or into more fractions depending on the number of sieves and the openings provided. The grains on the tray move down in a circular motion and along the slope of the tray. The head rice flows over and gets discharged at the end of the slope. The broken rice fractions collected below the sieves are discharged separately. This method, however, has a major disadvantage in that the broken rice rapidly clogs the holes in the sieves. A device is sometimes provided that causes bouncing of rubber balls provided in the space in between the sieves. This cleans the sieves, although not perfectly.

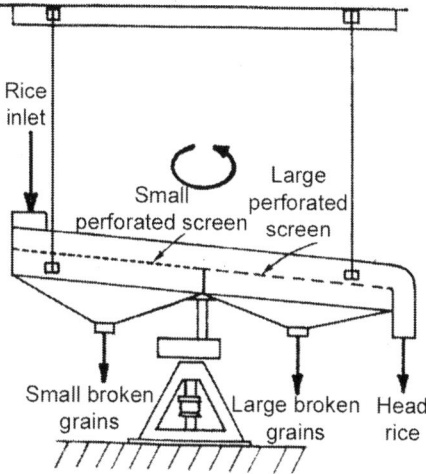

Figure 5.29 Plansifter for grading of rice [*Source*: Araullo et al. 1976]

Trieur or indented cylinder

This machine is also called as drum grader. As stated earlier, the machine was developed in France in 1845 for use in the seed industry. However, it was later inducted in rice processing first in Spain and has stayed since then as an important rice-grading machine and is being used all over the world.

It consists of a revolving cylinder with indentations, like small pockets, on its inner wall. The cylinder is set horizontally with its feed hopper side slightly elevated (Fig. 5.30). The rice (head rice and broken rice) is fed into it at the raised-end of the rotating cylinder. As the mixture moves slowly downwards towards the distal lower end of the rotating cylinder, the broken grains fall into the indents and sit nicely in the pockets of the rotating cylinder. Head rice, with longer grain length, cannot sit into the small pockets, and continues to slide and move down the cylinder length as it rotates. The broken grains which are caught by the cylinder inside the pocket continue to remain inside as the cylinder rotates upwards. They finally fall out of the pockets, at a higher point, by gravity, as the indent becomes inverted and unable to hold the kernel. As they fall, they are collected by a collecting tray placed centrally inside, along the length of the cylinder. The separated streams are then discharged separately at the end of the cylinder.

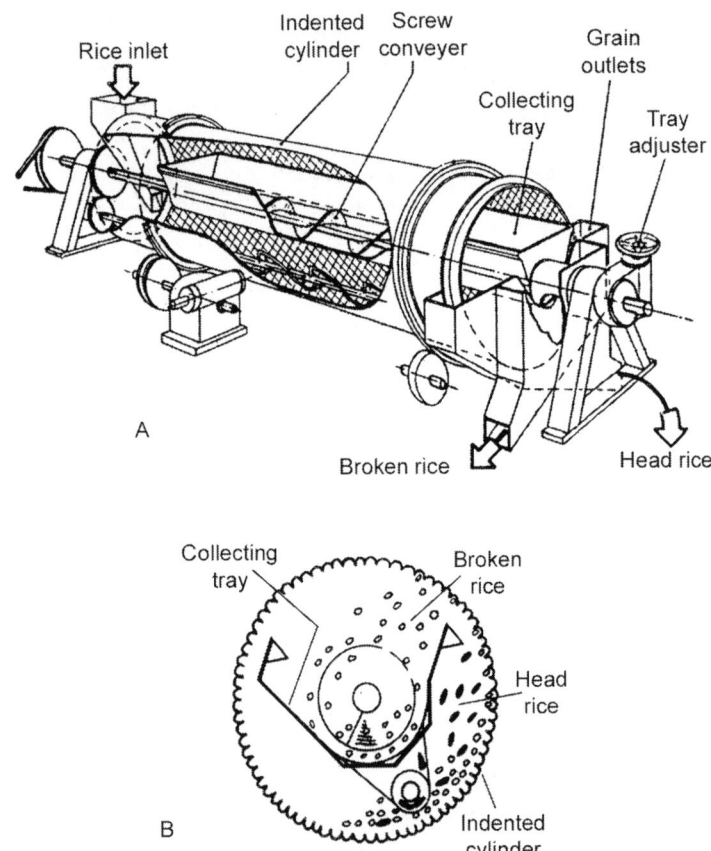

Figure 5.30 Trieur or indented cylinder. (A – unit showing part details [*Source*: Borasio and Gariboldi 1957; reproduced with permission from FAO]; (B – cross section of cylinder showing internal arrangements [*Source*: Araullo et al. 1976. Fig. 31]

By adjusting the size of the indentations and the position of the catch trough, broken rice of different sizes can be separated. As it separates the grains based on length, it can also be used in all seed-processing lines. For example, even in rice-processing systems, it is used in Country Elevators in

Japan, to eliminate the brown rice mixed in paddy.

Indented disc separator

The machine (Fig. 5.31) consists of a number of cast-iron discs, about 50 cm in diameter, mounted on a horizontal shaft at regular intervals inside a cylindrical case. Both sides of the discs have a number of indentations, radially arranged, to carry the broken rice. The discs rotate through a bed of rice in the bottom of a U-shaped trough. Depending on the size of the indent pocket, the broken grains are picked up as the disc turns through the rice. The broken grains remain in the pocket until turning of the disc results in an upside down position, when they fall out into the small collecting trough, provided in the centre, and are discharged into a collecting tray. A screw conveyer provided in the tray moves the broken grains to an outlet port. The longer grains (head rice) are conveyed forward by the conveying flights in the drum to the discharge end.

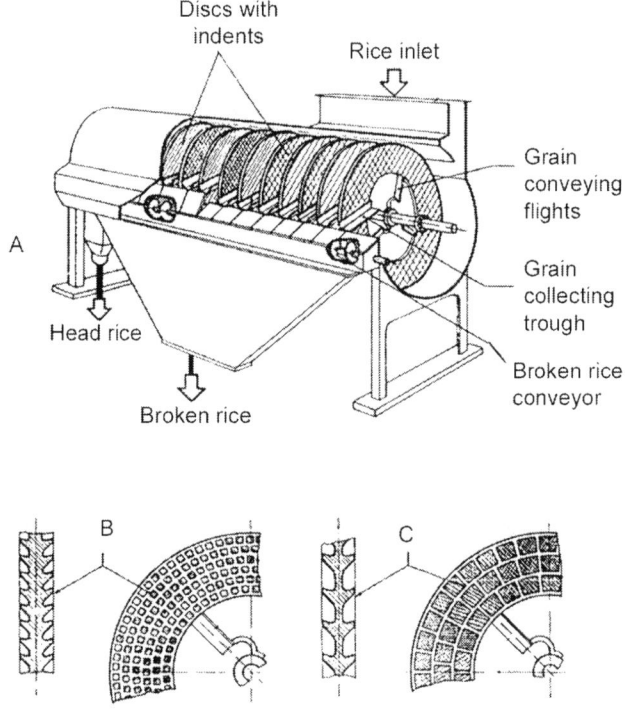

Figure 5.31 Indented disc separator (A – unit showing some internal structures; B and C – parts of the discs with small and large indents respectively with cross-sections of the discs) [*Source*: Borasio and Gariboldi 1957; reproduced with permission from FAO]

The advantages of this separator are its minimal clogging and little empty volume. This machine is employed in some of the American and Japanese rice mills and is not so common in the Southeast Asian countries.

Colour sorter/grader

The colour sorter machine is an optical sorter. It is used for removal of discoloured grains and foreign matter of the same size as the rice, from which broken grains have already been removed.

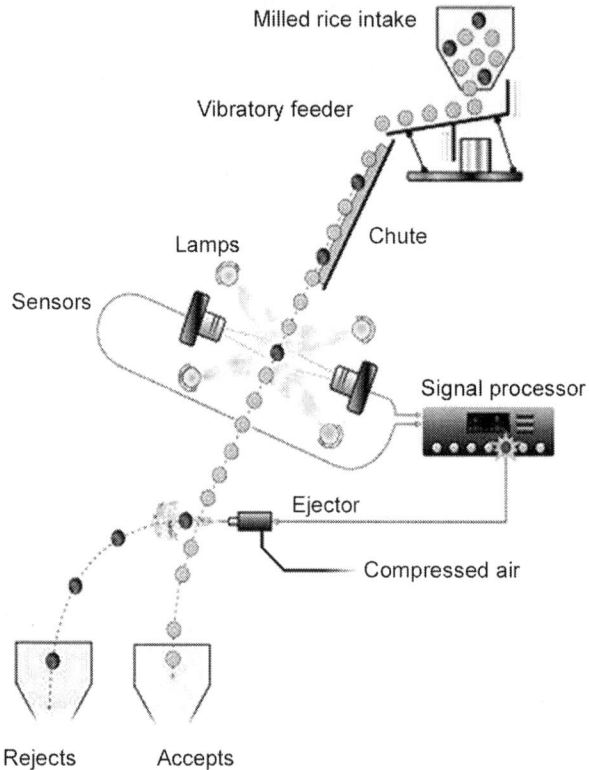

Figure 5.32 Colour sorter/grader. Reproduced, with permission, from Satake Group, Japan

In this machine, rice is supplied in an oriented manner, by chutes as guide channel, to pass through a photo-detector (sensor) system, one grain at a time (Fig. 5.32). The sensor detects the light deflected from the grain surface and compares it with the standard colour (background), which is adjustable. The grains that differ from the standard colour are blown out by a strong jet of pressurised air from the ejector. In the recent machines, it is possible to set the colour to any desired shade, thus making it possible to get uniform colour of the product. Although the detection and the response time for the activation of rejection of grain is in milliseconds, some good grains also get ejected in the process to cover all the discoloured grains. The rejects in such cases are recycled through the machine to bring down the content of the reject fraction. The capacity of the machine depends on the number of channels. Each channel normally has a capacity of 60–80 kg/h.

Till recently, use of these machines in most of the Southeast Asian countries was limited to rice mills engaged in export of high-value rice. However, they are now being increasingly adopted in rice mills for domestic markets also, as the demand for quality is increasing in domestic markets too.

5.2.10   Weighing and bagging

During the flow of material from the beginning to end of the processing, automatic weighing scales with appropriate containers are installed at strategic points. Their function is to receive material (paddy/rice/ fractions) in a continuous flow and release it in exact predetermined quantities by weight. At the same time, the number of weighments made is also recorded by an automatic counter, thus giving an accurate record of the total weight of the material processed in the mill.

At the end of the of processing line, the products are finally bagged and packaged for further storage or shipment, by automatic bagging/packaging machines which ensure dust-free bagging and sealing/stitching of the unit package, while the number of weighments made is also automatically recorded. The bulk packing normally is done in plastic woven sacks or gunny bags. Unit packs of 1, 2, 5 or 10 kg are normally done in HDPE plastic bags, as consumer packs, with appropriate properties of strength and vapour permeability to last through the retail marketing and storage requirements.

5.3   Mill operation and maintenance

The mill operations begin with the arrival of paddy to the mill area. Even before this, the decision of the lot to be taken is done with the assessment of its quality based on various factors, like the content of organic and inorganic impurities, damaged kernels, purity of the variety, and most importantly, its moisture content. Paddy having moisture content more than 14% needs to be dried to 14% or less, before it is stored or taken for milling.

The mill layout should ensure a smooth flow of material, right from the paddy intake point, from one stage or machinery to the other, to the final stage of bagging and packing. The capacities of individual machinery should be suitably matched and their placement be given due attention to ensure optimal space utilization, while ensuring at the same time a continuous and smooth flow of material. Many manufacturers market different type of machinery for the same unit operation based on different scientific principles. Many a times, millers make a judicious combination of unit operation machinery on the basis of perceived performance and others' experience to set up their own rice mills. In this case too, it becomes extremely important that their capacities should be properly matched for the optimal flow of material. Once the paddy begins to flow, an appropriate flow rate is required to be maintained. Interruption in the flow of material, or an excessive flow, both cause problems. For example, it is normally suggested that the standard flow rate for the paddy separator should be slightly lower than the basic capacity, and that of the polishing machine slightly higher, although these may be influenced by the quality of paddy.

There are basically three check points in the mill operation. First, in the shelling operation it should be checked whether the paddy is getting correctly dehusked and the resultant brown rice line is free from paddy grains. Second, in the polishing stage, the degree of bran removal (degree of milling) and the grain breakage need to be checked, and the third, in the grading operation, the amount of broken grains being generated need to be monitored. Flow rate of the material, as well as the individual machine parameters, are adjusted according to the results of these check points.

Training of the workers in the mill for the assigned tasks by competent technical personnel is necessary for smooth functioning of the mill, including operation methods and quality control. Periodical check-up and servicing should be routinely carried out in order to prevent failure of the machinery and reduction in their capacities. Manufacturers' recommendations for maintenance of the machinery and equipment should be followed. It is better that the machinery supervisors/operators are trained by the machinery manufacturers for the optimal handling and monitoring of the respective unit operation.

Another very important activity for maintenance of the mill is its periodical thorough cleaning. Most of the rice producing countries fall in the tropical zone, which have climatic conditions that favour microbial, insect or pest infestation. Hygienic conditions in the mill area should be maintained to avoid infestation. Microbial infestation can be avoided by maintaining moisture content below 14% level for storage of paddy and rice for short periods, and below 12% for long periods. Pest infestation could be controlled by appropriate design of the mill building and its premises. However, control of insect infestation needs periodical measures to be taken. Elevator boots, corners and pockets formed in the machinery, and dead spaces created while conveying and handling the material, especially chutes and pipes, are the places where grains and their debris get accumulated. These provide perfect breeding grounds for the insect activity and become continuous source of infestation with eggs, larvae and live insects, inside the mill. A thorough pneumatic cleaning with high pressure air, and fumigation with the permissible insecticides, should therefore be resorted to periodically, for maintaining a good quality of the products coming out of the mill.

5.4   Recent developments in operation and control

Rice milling process has come a long way, over the past 200 years, from its primitive hand pounding system to a complex and sophisticated technology as it is being practised today. This has been possible by understanding the grain, its anatomical, physical and chemical properties on one hand, and the application of matching engineering principles on the other, for its optimal processing. And the technology is constantly getting upgraded. New technologies and new machineries are being introduced. Modernization is not confined to any fixed time frame. It is always the present, which is modern. There is no completion or termination of technology.

The past few decades particularly have seen a tremendous change in the way the rice grain is being handled and processed. Improvements have come with a remarkable speed. A brief glimpse of the basic machines involved at various stages of milling has been presented above. What must be mentioned additionally, however, is that the introduction of electronics into the rice milling system, and the automation which is being incorporated and integrated at almost all the steps of processing, has enabled accurate operation and control of each aspect of rice milling with economic advantage.

Large rice mills, employing all modern and state of art technologies, are already working in the advanced and industrialised countries. In fact, the world's first fully computerised rice mill with automatic processing controls was built about three decades ago, in 1981, by Satake Engineering Co. and installed in Kanagawa prefecture of Japan. A few large scale units are operating in other Asian countries too. These mills provide data on yield, operation, condition and performance of each machine, degree of polish, sales details, operating cost, profit, etc. at the flick of a button.

5.5   Effect of some grain parameters on milling results

Although there are a number of parameters which need to be considered and given due attention for monitoring the health and economy a rice mill, a few important indices are briefly discussed below that are commonly used in the context of milling, and output of rice and its by-products.

1. *Content of impurities, foreign matter, dockage etc.:* This is an important parameter. If this fraction is higher, the quantity of clean paddy, and that of the milled rice produced from it, gets naturally reduced. The intake paddy should therefore have the least amount of this fraction.

2. *Moisture content of paddy:* The moisture content of paddy for safe storage or milling should be less than 14%. Paddy with higher moisture content is prone to fungal infestation, development of hot spots and cake formation during storage. This increases the breakage of rice during milling and drastically reduces the quality of milled rice. For long storage, moisture content of not more than 12% is advisable. Further, it is also advisable to maintain uniform moisture content in the whole paddy mass being milled. Sufficient measures need to be taken to ensure this. Humidity in the mill area at the time of milling also affects the milling performance. If the atmosphere is dry the rice tends to lose moisture, while it would gain if the mill has high humidity. A loss or pick up of up of about 0.5% moisture by rice during milling is possible due to large difference in the humidity of the mill area and the moisture content of paddy/rice with which it can equilibrate.

3. *Husk content of the paddy:* This has a direct effect on the total output of rice from a given amount of paddy. Its content is a varietal characteristic. Most varieties have husk content in the range of 19–22%. In general, shorter and roundish grain varieties have lower, while long grain varieties have higher husk content.

4. *Degree of milling/degree of polish:* It is the extent to which bran is removed from brown rice during whitening and polishing process. It is, therefore, expressed as per cent in terms of brown rice. Normally, to give the milled rice a better marketing, storage quality and consumer appeal, a degree of milling in the range of 8–10% is practiced. It is some time measured indirectly in terms of 'whiteness' of milled rice in comparison to that of brown rice, by employing reflectance meters.

5. *Yield of milled rice / milling yield:* It is the total milled rice obtained after milling from a given amount of paddy, and is expressed as per cent in terms of clean paddy, from which the rice was obtained.

6. *Yield of head rice:* This is another important parameter, and the major contributor to the economics of rice milling. As in the case of total milled rice yield, this parameter is also normally expressed as per cent in terms of the paddy from which it was produced.

7. *Yield of broken rice:* It is the content of total broken grains in a given amount of milled rice, and expressed as per cent of total milled rice. Normally, broken grains having a length up to three-fourth of the whole grain are classified as broken rice. The rest, consisting of whole grains, and broken grains having length of three-fourth of whole grain, and above, are classified as 'head rice'.

Further reading

The subject of rice milling has been discussed and reviewed extensively in:

- Gariboldi (1974)
- Wimberley (1983)
- van Ruiten (1985)
- Satake (1990)
- Wadsworth (1991)
- Food Agency (1995)
- Bond (2004)

6

Ageing of rice and how to promote or retard it

Cereal crops including rice are grown at best once or twice in a year. But being the staple food of humankind, they are required all the year round. Therefore, cereal grains have to undergo long storage for up to a year or more. The character of cereal grains too is suitable for their long storage. As explained in Chapters 3 and 4, food grains tend to be in equilibrium with the surrounding atmosphere. It means food grains are low-moisture food and hence can be stored for long periods without appreciable deterioration. This ability of cereals being stored for long periods is of fundamental importance to human civilisation, being the basis of providing food security to humankind.

6.1   Change in the cooking–eating properties of rice during storage

Rice is unique among cereals in more ways than one. Not the least of these ways is the response of rice to storage. All food grains must be stored. The chief concern during this storage is how to adopt such ways and means as would prevent qualitative or quantitative deterioration of the grain during its storage. But in the case of rice, there is an additional matter of importance. Rice is unique among food grains in that it undergoes a clear and dramatic change in its cooking and eating behaviour with progress of time during its storage. Soon after harvest, rice tends to cook generally to a rather sticky, lumpy and moist mass. This character gradually changes over time. Finally, after several months of storage, the rice now cooks comparatively more dry, fluffy and free-flowing. So rice stored for some time (generally called 'old' rice in south Asia) is not quite the same as rice recently harvested (generally called 'new' rice in south Asia). And this phenomenon of change in the cooking–eating behaviour of rice with time after harvest is called the ageing of rice.

6.1.1   Difference in preference for 'new' or 'old' rice

Interestingly, different people respond differently to the change in the cooking property of rice during storage. People in south Asia (the Indian subcontinent) heartily dislike the cooking behaviour of new rice and look forward to its ageing. In fact there is a substantial price differential between new and old rice and people are happy to pay a higher price for the old rice. This consumer behaviour is also true of west Asia (Middle East). But one would be wrong to think that this is a universal response of rice eaters. People in the east and northeast region of Asia (Japan, Korea, and northern China) show a diametrically opposite preference. They like new rice and have a hearty dislike for old rice, which they consider to yield too hard, individual and 'stale' cooked rice.

This difference in preference for new and old rice between the two regions (south and east Asia) may have a reason. It may have something to do with the difference in the type of rice that is naturally grown in the respective regions. Indica-type rice is grown in the Indian subcontinent. This type of rice by nature cooks comparatively somewhat hard, nonsticky and dry. People in this region may perhaps for this reason be inclined to prefer old rice rather than new rice, for old rice reinforces these characters. Japonica-type rice, which by nature cooks sticky, lumpy and moist, on the other hand, is what is grown in east Asia region. Consumers are therefore naturally habituated to this taste. That may be the reason why they like new rice and dislike old rice.

Be that as it may, a substantial and often dramatic change in the cooking–eating behaviour with time of storage, i.e. ageing is an inherent property in the case of rice. This is also a unique property of rice and has not so far been found in other food grains (such as wheat, maize, sorghum, millets, legumes).

6.1.2   Ageing changes as measured in the laboratory

Scientists, being also consumers, were well familiar with this unique phenomenon of ageing of rice. So they naturally became interested to study it. This phenomenon was first studied, so far as is known, by scientists in Bangalore in India (in the Central College and in the Indian Institute of Science) from 1930s. The work was subsequently taken up in many other countries as well from 1950s onwards (Japan, USA, Spain). First they studied and measured the precise nature and extent of the changes in cooking–eating properties that consumers perceived to happen in rice up on storage. Later they tried to investigate the changes in the subtle chemical constituents and other properties of rice in the hope of explaining the mechanism of the gross changes in cooking–eating behaviour.

The quantitative changes in the cooking–eating behaviour of rice with time of storage that they observed were:

(a) The hydration properties of the rice (i.e. the amount of water absorbed by the rice during cooking in a given time, or the rate at which it absorbed water during cooking) changed progressively with time after harvest. Initially, there was an increase in the rate of water absorption for a few months. But the rate then steadily declined with time of storage and this decline went on for years. Thus the long-term trend was a steadily decreasing rate of hydration with time of storage.

(b) The new rice lost a substantial amount of grain solids during cooking (6–10%) into the cooking water. This loss progressively decreased with time as the rice aged. The excess cooking water, if any, for this reason was thick and viscous in the case of new rice. But it became thin and freer flowing after ageing.

(c) New rice grains were rather soft, moist, sticky and lumpy (grains clinging together) upon cooking. But old rice grains cooked perceptively different. This was confirmed by precise measurements with laboratory instruments. Cooked rice grains thus became progressively dryer, harder, more free-flowing and fluffy as the grain aged.

(d) New rice when cooked swelled but little in the cooking pot. Old rice on the other hand progressively swelled more freely and looked fluffy in the pot. This too could be shown by precise laboratory procedures.

(e) Rice grains elongated (became longer) up on cooking. Interestingly, new rice grains elongated less, while old grains elongated very well.

It was thus confirmed that the above changes were not simply perceptions of consumers but could be demonstrated also by sophisticated laboratory measurements. Some of these differences are illustrated in Fig. 6.1.

Other changes that were observed and measured by laboratory systems were:

(a) Rice has a specific paste property. If rice grains are powdered, dispersed in water and the mixture is gradually heated under specified laboratory conditions, then a paste is formed. These pastes have a definite behaviour pattern. But the pattern of rice pastes went on changing as the rice aged. Thus:

• the peak viscosity of the paste at any given concentration showed a long- term trend of decrease with time as the rice aged,

• the paste characteristically thins down ('breaks down') when it is heated with stirring. But this 'break down' of rice paste decreased as the rice became old. That is, the paste became progressively more and more strong and stable as the rice aged, and

• as the rice grain became harder (when cooked) upon ageing, so the cold rice paste too showed an increasing viscosity as the rice aged.

This changing paste behaviour suggested as if the starch granules themselves in the rice grain seemed to become progressively more and more strong and resilient and so resisted disintegration as rice aged.

(b) There was a progressive decrease in the solubility of the grain substance (grain solids, amylose starch, protein) upon cooking of rice or its powder in water as the rice aged.

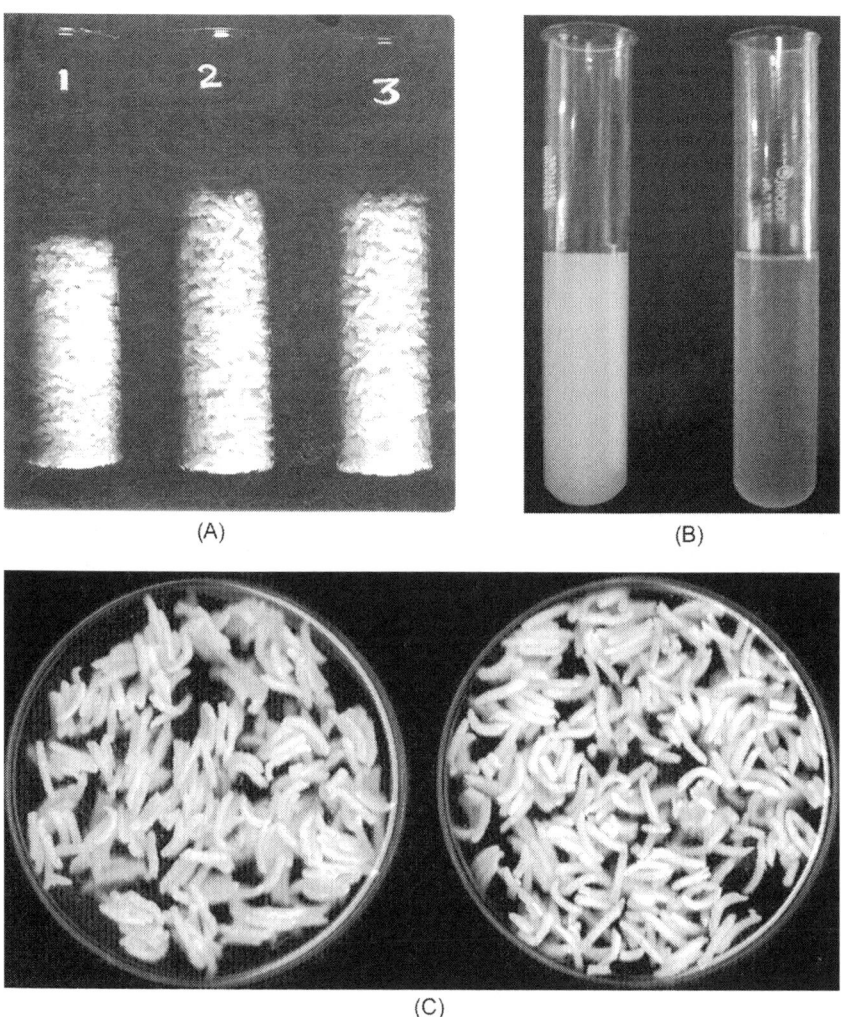

(A)

(B)

(C)

Figure 6.1 Differences between cooked 'new' (freshly harvested) and 'old' (aged) rice: (A) New rice (tube 1) shows poor volume expansion during cooking as compared to old (tube 2) or artificially aged (tube 3) rice. (B) New rice (left) loses more solids in excess cooking water compared to old rice (right) [Reproduced, with permission, from Bhattacharya (2011)]. (C) Old rice grains (right) elongate more and cling together less upon cooking compared to new rice grains (left). [*Photo courtesy:* Charles Stephen]

Taking these observations together, a picture that emerged was this: It seemed that the rice grain or the grain substance, might be its starch granule itself, were rather weak or loosely organised when new. So they hydrated and swelled easily, and also then disintegrated relatively easily. But they became progressively more organised, reinforced and strengthened as they aged. They did not now hydrate as easily as before. The granules did not swell, nor disintegrate as easily as they did soon after harvest. And the extent of these changes increased progressively with time of ageing. Simultaneously, in agreement, the solubility of the grain constituents (amylose starch, protein) too declined progressively with time. Also, the grain as a whole became progressively hard (upon cooking) with reduced loss of grain solids during cooking as it aged.

6.2   Theories of rice ageing

Scientists have thus meticulously collected and recorded information about the precise changes in various properties and constituents of rice during its storage. Yet the phenomenon of its ageing is still a mystery. What caused this overall change in cooking–eating properties? Research so far has shown, as summarised in the previous paragraphs, that the rice grain or the grain substance progressively became more organised, reinforced and resilient over time during storage after harvest. But why and how did this happen? What natural force or basic factors brought out this overall transformation? Here unfortunately our understanding has not grown proportionately. A number of scientists in several laboratories all over the world have spent a lot of time and efforts to understand or to clarify the underlying basis of the rice ageing process. Lot of information have been collected; many changes in starch, protein, lipids (fat) and carbohydrates have been recorded and measured. But a clear picture of how all these changes add up to explain the overall process of rice ageing has not yet emerged. Besides, one must remember, no other cereal grain has so far been reported to undergo a similar process of ageing. This fact is another surprise. Barring small differences in contents of various constituents and some properties, all cereal grains broadly speaking are basically similar in nature. Therefore, why so prominent a process like ageing operates in the rice grain but not in other cereal grains is another matter that remains to be explained.

Be that as it may, we can perhaps broadly mention the various theories that have been offered to explain the process of rice ageing. These are:

(a) One theory proposed that starch gets partially polymerised during storage through residual enzyme action, which may explain the altered behaviour of starch.

(b) According to another theory, the principal protein of rice, oryzenin, undergoes polymerisation probably partly through disulphide bonding, which is what causes the changes.

(c) Sulphydril-disulphide transformation has undoubtedly been shown to happen in rice protein during storage. The resulting bridge formation has been considered to be a primary factor behind the ageing phenomenon.

(d) Other theories suggest that lipids play a major role. One theory suggests that free fatty acids (FFA) develop from fat. This FFA associates with starch, and this is what causes the progressive transformation.

(e) Another suggestion, on the other hand, is that lipids undergo autoxidation and forms carbonyl compounds. These carbonyls cause cross-linking with starch and protein bodies leading to gradual strengthening of the grain substance.

(f) Finally there is a theory that the ageing process is mainly related to non-starch polysaccharides. The cell wall is formed by non-starch polysaccharides along with cellulose and other compounds such as ferulic acid. These cell walls in rice are said to gradually harden over time through cross-linking and then to hold the grain together.

Bits of evidences in favour of any or all of these theories have been collected. But none of these unfortunately succeeds in explaining the overall process. Either there are contradictory evidences or some vital links are missing. Or no possible justification is available to explain why the same system does not operate in other cereal grains. Overall one can say that ageing of rice still remains an enigma despite strenuous efforts over decades of research by a number of workers in different laboratories. Ageing of rice perhaps remains the last frontier of rice-grain research.

6.2.1   Effect of temperature on ageing of rice

One characteristic of the rice-ageing process should be noted here. It has been repeatedly observed that the prevailing temperature strongly affects the rate of ageing of new rice after harvest at any point of time. The higher the temperature at which the rice is placed, the faster does it age and vice versa. Thus if rice is stored cold, for instance at 0°C or −5°C, then that rice may remain virtually new for a long time and may require years to even partially age. On the other hand, if the rice is stored at a high temperature (say 30°C or 40°C), it ages quite fast. The rate of ageing is always proportional to the temperature. It may be noted that this relationship applies not only to rice immediately after harvest but at any stage thereafter. Thus, if the rice is already partially aged, let us say at 20°C and is now shifted to 40°C, the increased rate of ageing still operates.

This relationship by itself may not be considered as something very novel. Increased temperature always increases the rate of any reaction. But the above case is of interest from two angles. One is because of its practical value. Thus the Japanese use this relationship to delay rice ageing by putting rice in cold storage (see later). Similarly, Indian exporters of basmati rice to west Asia often ship the new rice to Dubai and store it in the hotter weather there for quicker ageing than in the cold winter of north India. The other point of interest is that this temperature dependence of the ageing process applies not only to natural ageing but also to artificial ageing by application of heat treatment (see later). This fact raises questions about the nature of the ageing process.

6.3   Is it possible to intervene in the process of rice ageing?

Ageing of rice is a natural process. It happens whether one likes or dislikes it. But there are strong likes and dislikes associated with the process. People of south Asia (Indian subcontinent) wait eagerly for rice to age. Those of east Asia (Japan, Korea, northern China) strongly wish to prevent it. So with so much strong emotions involved with it, scientists engaged in these studies sooner or later started asking: Can one intervene in the process of rice ageing? In other words, can one either promote (hasten) or prevent (or delay) the process?

6.3.1   Artificial ageing of new rice: Principles

In the background of the strong dislike of new rice in India and the pioneering work of Indian scientists to try to understand the process, this question was raised in India more than half a century ago. Scientists in the Central Food Technological Research Institute (CFTRI) at Mysore in south India took up the matter. They wondered whether any treatment could be given to the rice that would make new rice to cook like old rice. In other words, they wanted to study whether the process of ageing could be hastened or accelerated by an artificial means. The idea came from their observation of a system that rice farmers sometimes adopted. Rice in stalk was sometimes stacked in the field and left aside for some weeks to dry. This process caused some internal heating in the stack. It also caused some discolouration of rice ('stack burning'). But what was striking, such rice usually cooked a little better than usual new rice. This observation led to the idea that perhaps heating could be the answer.

Steam treatment

Accordingly, it was found that if paddy after harvest was steamed and kept hot for some time, the resulting rice milled from the paddy cooked rather like old rice. A process of steam treatment of new paddy was developed in the CFTRI on this basis in the mid-1950s. It was called 'curing' or steam curing of rice. The process of course had some shortcomings. One, the steamed (and dried) paddy broke more than normally when it was eventually milled. In other words, its head rice recovery was affected. Secondly, the milled rice also looked different from normal. It looked a bit like chalky and somewhat opaque. These were its defects. But its efficacy for improved cooking was without any doubt. Gradually the process became very popular first in and around Mysore city, later gradually in the Karnataka state, and eventually after decades now throughout the country. This despite its unquestioned defects mentioned, because to Indians poor cooking of new rice is a much more serious factor than grain breakage and appearance (opacity). The product is popularly known as 'steam rice'. It is now quite popular even in north India, including for basmati rice in the domestic market (but not for export, where the altered appearance of milled rice would be a serious handicap).

Dry-heat process

Scientists in CFTRI at Mysore went further. Considering the deficiencies of steam curing mentioned above (higher breakage and grain opacity), they wondered whether milled new rice instead of paddy could be directly cured (so as to avoid the additional milling breakage). Milled rice unlike paddy could not be steamed. So they tried a process of dry heating. But dry heating of rice could not be done in an open vessel, for the rice was totally damaged by drying and cracking. So, small amounts of milled rice were taken in closed containers and heated from outside. Different temperatures (40–100°C) and different relative humidities (50–90% RH) were tried for the treatment. It was found that heating under appropriate conditions indeed led to excellent curing of new rice. The degree of curing was strongly dependent on the temperature. The higher the temperature, the faster the new rice got cured, and vice versa. What is more, the rice in this case appeared almost indistinguishable in appearance from untreated but aged milled rice. The process thus looked very promising. Unfortunately, the process, simple in theory, was operationally very complicated. Therefore it could not be scaled up and remained merely a theoretical possibility on paper for almost 40 years. Ultimately, the matter was taken up in a rice company later and a successful large-scale process was developed. This was called the process for accelerated ageing of rice (AAR process).

These two processes for artificial ageing of rice are now described below.

6.4   Technologies for accelerated ageing of rice

6.4.1   Steam curing

The steam curing process is really very simple. As long as its deficiencies are accepted, it is a simple and effective process for imparting the cooking–eating behaviour of old rice to new rice. All it need is to steam the paddy, keep it hot for some time and finally dry it to remove the moisture absorbed during steaming. The steaming per se can be carried out in various ways. Although discovered more than half a century ago and by now so fairly popular in India, there is not much of a standard process that has emerged. Millers do the steaming in whatever systems that is available. The basic requirement is to have a suitable vessel to hold the new paddy and to pass steam into it from a boiler.

Since most of middle- to large-scale rice millers in India also run a parboiling plant, the basic facilities are already available with them. Most working parboiling plants in India use the rather primitive 'double-boiling' parboiling system or its modernised version (see their description in Chapter 7). In this system, paddy is taken in small hopper-bottom kettles (capacity 200–500 kg paddy) and steam is passed into the paddy through a vertical open pipe. The paddy thus steamed is dumped into a big tank placed below. Paddy is thus continuously steamed in one kettle after another until the tank below is full. Water is then run in for the paddy to be soaked.

With this facility already available, millers use the same or similar facility for the steam-curing process as well. If spare steaming kettles are available, then these are used for steaming. If not, more are bought. Or else bigger steaming tanks holding 1–3 metric tonnes of paddy are purchased from equipment suppliers and used. In the latter case, a simple vertical pipe may not be enough for steaming. Instead a rudimentary steaming manifold with one main and several branch pipes is employed. The pipes are closed at the end but have perforations for steaming (Fig. 6.2).

So the basic process consists of taking paddy into the steaming tank and passing steam into it. With multiple small-capacity steaming kettles, a relatively low-capacity boiler (1–2 kg/sq cm, gauge) would be sufficient. But with larger steaming tanks, the boiler capacity has to be commensurately increased. This is to enable the steaming to be completed in a reasonable time (10–20 min). Once steamed, the paddy has to be kept hot either within the same steaming tank or in a separate tank or in a heap on a platform.

The time of hot storage has to be adjusted from two angles. The main factor is how fresh or old the paddy is. Fresh paddy needs rather prolonged heat treatment to cure. The required time decreases as it becomes older. Thus the time of hot storage would vary from perhaps 30–75 min at the beginning of the season (when the paddy is fresh after harvest). It may be perhaps 10–30 min six months after. But there is also a second factor: the eventual product colour. The applied heat imparts not only curing effect on the rice but also a faint to light browning effect. The longer the heating, the more pronounced are both the effects. So the miller has to adjust the system to get the best curing with least discolouration of the milled rice.

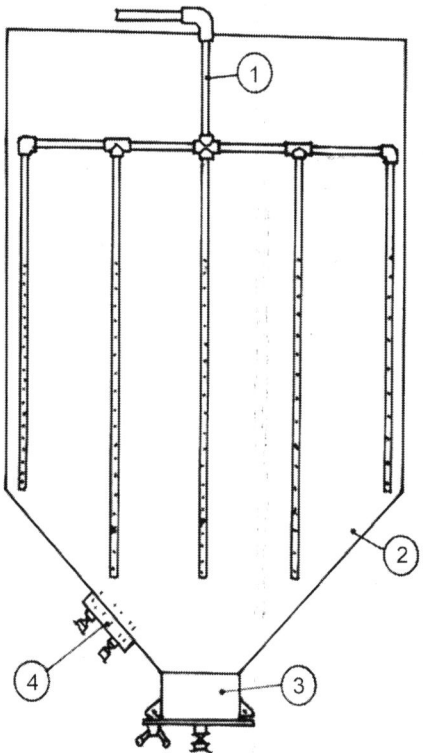

Figure 6.2 Schematic drawing of tank with steaming manifold for steam treatment of paddy

The moisture content of the paddy at the time of steaming is another important factor to be kept in mind. Too high paddy moisture during steaming (>14.5%) imparts a superficial parboiling effect to the rice. That is, the rice after milling has a faint, though distinct, surface translucence and light brown colour and also a slight oily appearance. As against this, grain breakage during milling would be relatively a bit low. Although Indians are by and large more concerned with the cooking behaviour than with the appearance of their rice, this slight superficial parboiling effect is generally not liked by those who eat raw white rice. Too low moisture in the paddy (<13% moisture), on the other hand, tends to create an opposite effect: grain opacity. The milled rice acquires a slight chalky-like appearance in addition to becoming rather more prone to grain breakage. So the ideal moisture content is 13–14% moisture in the paddy, which gives the best compromise. To summarise, both paddy moisture and steaming (or hot-storage) time need to be so adjusted as to get the best compromise of curing effect, grain breakage and grain appearance and colour.

The paddy absorbs some moisture during steaming, amounting to about 3–5%. So the steamed paddy after hot storage has to be dried. The drying should be carried out under as mild a condition as feasible. For example, drying could be done with 45–55°C air in an LSU dryer or in the yard in relatively mild sun, preferably in two stages with tempering in between (especially if the moisture is on the higher side). As already explained, even then the paddy yields 4–5% extra broken rice because of the steam process. But traditionally, in the general Indian market, Indians are less concerned with broken rice or grain appearance than with the cooking–eating behaviour of their rice. So some extra broken rice, within limits, would pass in the general Indian market. Of course, the situation would be different for high-value markets. For example, in north India, high-value basmati rice is being steam-processed these days (for the domestic market, not for export). But in this case, broken grains are generally separated and marketed separately. The same situation may prevail whenever a high-value product or a branded product is marketed perhaps in a supermarket.

Steam curing at home

Interestingly scientists in the CFTRI also devised an ingenious system for domestic curing of new rice. Although the system does not appear to have become popular, it is briefly described here to show the powerful effect of steam.

A special rice cooking vessel is central to the system. It is a simple, upright, cylindrical pot with a lid, in which water is first boiled. Rice is not taken directly in this pot. Rice is taken in a well-fitting, perforated circular tray that can be hung within the pot. The tray is first hung above the water level in the cooking pot and the rice is allowed to be steamed from the water boiling below. Steam rising from the boiling water passes through the perforations into the rice taken in the tray. After adequate steaming, the tray is now lowered into the boiling water and the rice is allowed to cook under cover. Because of the steaming, the rice, even though new, cooks remarkably well as per the Indian taste.

6.4.2   Accelerated ageing of rice by dry heat: AAR process

As tested in the laboratory-scale process mentioned earlier, the dry-heat artificial ageing of rice involves three essential steps:

(a) The rice is first to be heated to a high temperature (generally 85–90°C),

(b) Next there should be a system to store the heated rice hot without losing its temperature for as long as it takes for the curing process to be completed (1–12 hours), and

(c) Finally, the hot rice has to be cooled to ambient temperature.

These three steps in the process appear simple in principle but are very complicated in practice because of certain specific properties of rice and the customary way it is consumed. Rice is overwhelmingly consumed in the whole-grain form, not as flour. Any action or process that affects the grain integrity or its breakage during milling is therefore unacceptable. It is primarily here that the problem of dry-heat system lies.

First, as explained in Chapters 3 and 5 on drying and milling, rice is mechanically strong and its integrity is not easily affected by mechanical forces. At the same time it is extraordinarily sensitive to moisture change. Any process that tends to change its moisture content at a rather rapid rate (rapid hydration or dehydration) makes it susceptible to fissuring or cracking. And this cracking then is liable to cause grain breakage during milling or handling. One can see that all the three steps mentioned above, involving heating and cooling, are intrinsically such as to promote grain fissuring. So here lies the difficulty. However, the answer lies in the cause itself. If the heating or hot storage or cooling can be performed in a nearly hermetically sealed condition that avoids any loss or gain of moisture, then by definition it should be possible to avoid grain fissuring. In other words, for a successful AAR process, all the three steps mentioned above have to be performed such that there is only heat transfer but no (or negligible) mass transfer.

Second, another factor is water condensation. The EMC-ERH curve (see Chapter 3, Fig. 3.1) of rice is temperature dependent. The curve shifts slightly upwards with rising temperature, so that the ERH shifts upwards as the temperature increases. What it means is, increasing temperature affects the grain in a way as if the moisture in the grain tends to be pushed out as the grain temperature increases. Clearly, the walls of all the process equipment must be maintained at or near the temperature to which the rice is heated. Otherwise, if the equipment walls including all connecting chutes and pipes were at a lower temperature, there is bound to be water condensation at the walls. Any such condensation has to be avoided at any cost, for any contact between hot rice and liquid water would lead to its cooking, discolouration and spoilage. This determines the second condition for the dry-heat AAR process.

Third, the extent of curing the new rice undergoes is dependent not only on temperature but also on the time of heating. In other words, if any grain-to-grain variation in product quality is to be avoided, then all the grains have to be heated not only to the same temperature but also for exactly the same time. What this condition implies is that a batch system for the process is ruled out. A continuous system is necessary. A continuous system implies that the hot-storage step must be carried out in a moving first-in first-out (FIFO) system so that all the grains in the mass are exposed to heat precisely for the same time duration. FIFO operation is also needed, for any hold up anywhere would be liable to lead to heat damage and browning.

Fourth, the cooling step too has to be performed such that there is neither any condensation (meaning no contact-heat-transfer cooling) nor too much moisture loss in cooling by blowing air.

To summarise, for the dry-heat AAR process to be successful, the process has to be such as to:

(i)   be carried out in a virtually hermetically sealed equipment system so as not to lose any moisture,

(ii) have the equipment walls being maintained at a temperature roughly identical to that of the hot rice, so as not to have any water condensation thereon,

(iii) be conducted in a continuous-flow process with strict FIFO operation, and

(iv) not lose excessive moisture when the hot paddy is finally cooled by air.

Once understood, it is not that difficult to provide for all these conditions. The three steps of the process to which the above conditions are to be applied can be summarised as:

• First the rice is to be heated in a continuous, agitated, contact heat-transfer system.

• Second, the hot rice should be allowed to pass through a mass-flow FIFO bin such that all grains come out after a predetermined time of hot storage.

• Finally, the hot rice is to be cooled by humidified air in an efficient air-mixing system.

All the above conditions are well provided in the AAR process that finally emerged in the rice company where the scaling up of the process was taken up.

The process is shown in the flow diagram in Fig. 6.3. The three equipments are arranged vertically in three tiers so as to have all rice movements by gravity. Paddy is fed into the system at a carefully set rate through a rotary valve at the top. A Nara Paddle Dryer (NPD) (Fig. 6.4), with its curved and extended contact-surface paddles for heat transfer, is used for the heating step. Steam is passed through the hollow paddles at a pressure of approximately 1.5–2 kg/sq cm (gauge). With its extended heat-transfer surface of the moving paddles which keep the rice well agitated, almost as if in convection, the NPD can efficiently heat the granular grains to the desired temperature at a constant rate. The walls of the NPD and all downstream equipment including all connecting chutes and pipes are kept adequately heated either by steam tracings or by thermopads or at least efficiently insulated to prevent heat loss.

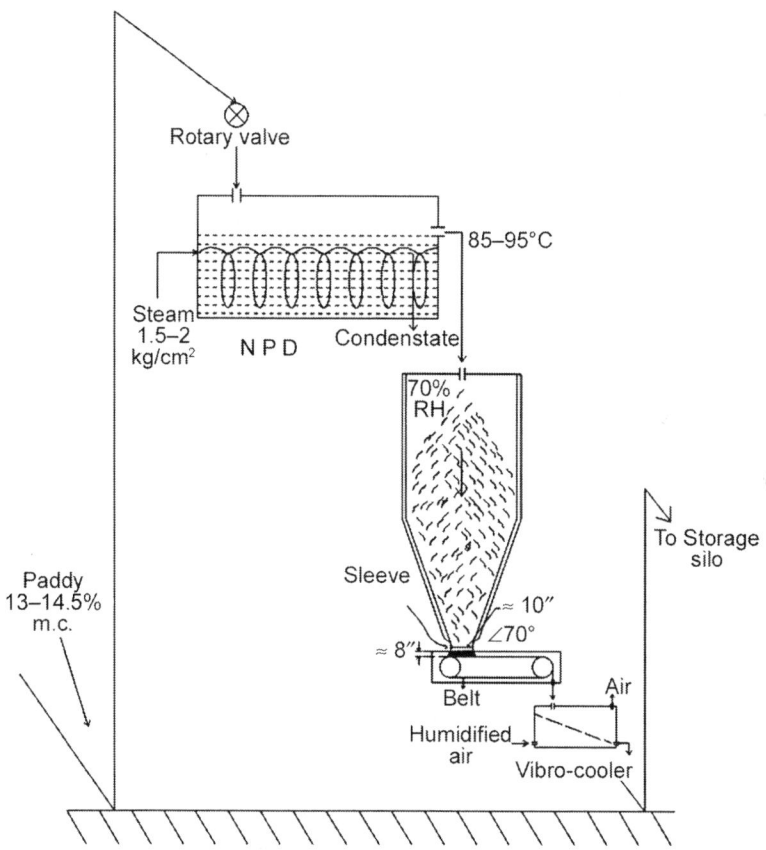

Figure 6.3 Flow diagram of dry-heat AAR process for accelerated ageing of new rice (not to scale) [Reproduced, with permission, from Bhattacharya (2013a)]

Structure of Nara Paddle Dryer (NPD)

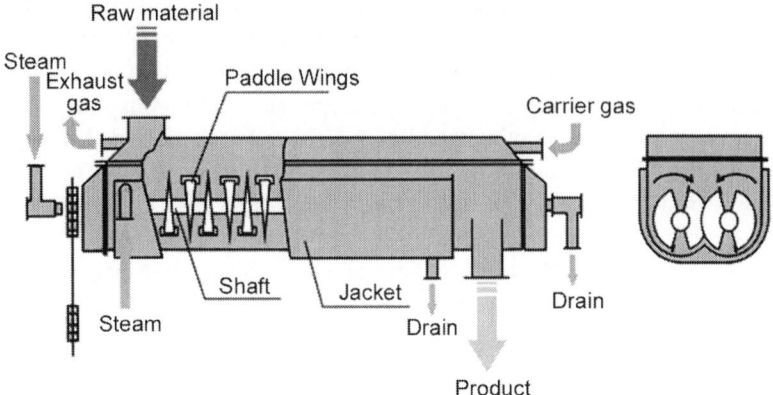

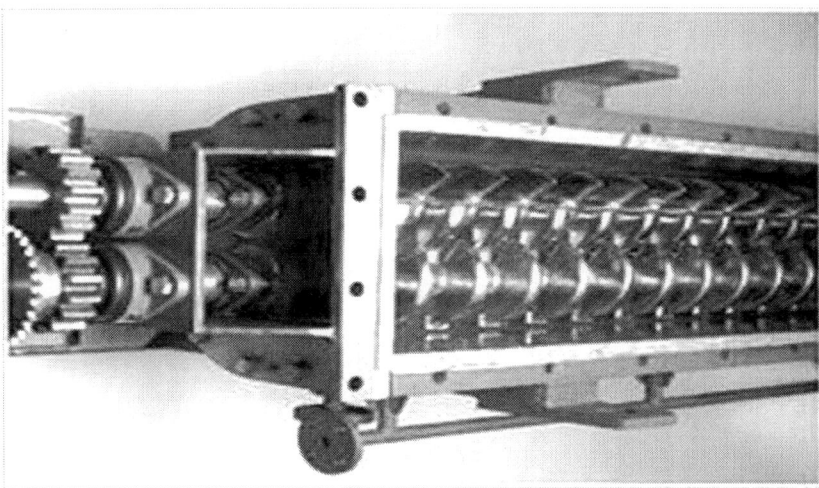

Figure 6.4 Nara Paddle Dryer (NPD) (Top: Schematic drawing, bottom: photo)
[*Courtesy:* Kilburn Engineering Ltd.]

The hot storage is performed in a steam-jacketed, stainless-steel, mass-flow bin having a holding capacity of 5–6 times the working capacity of the plant (say 30 tonne [t] holding capacity for a working capacity of 5 t/h). Steam is passed in the jacket at 0.3–0.5 kg/sq cm, gauge. The bin is strictly designed for mass flow (FIFO) with a hopper angle of minimum 70° to the horizontal and a bottom opening of 10–12 in. diameter. To this bottom opening is mounted a sleeve having an angle of 2° to the horizontal with the elevated side turned towards the discharge end of the belt below. The relative humidity in the bin is maintained at about 75% RH with an appropriate sensor and electronically controlled steam injector.

The hot paddy coming out of the bin is delivered to a moving belt made of appropriate heat-resistant material and placed 6–8 in. below the discharge opening. The whole belt system is enclosed in an insulated box. The movement of the belt and the sleeve position are carefully adjusted to deliver the paddy at an identical rate as the input to the NPD (as adjusted by the rotary valve).

The cooling unit has a slanted, vibrating stainless steel perforated plate enclosed in a box in which the hot paddy falls. The ambient air humidified by fine sprays of water is blown through the mesh to mix efficiently with the falling cascade of paddy such that the paddy is quickly cooled. The escaping warm air is passed through a cyclone to catch the dust and debris. The cooled processed paddy empties into a hopper and then into a collecting elevator.

It will be noticed that the system is almost self-operating with minimum controls. The rate of paddy feed to be processed is adjusted as per the required capacity by the rotary valve on top. Then the input of steam into the paddles of NPD is adjusted to get the desired rice temperature. The precise time of hot storage in the bin is achieved by simply adjusting the time of the first outflow of paddy from the bin. That is, the first withdrawal from the bin is started exactly after the required time of hot storage. Thus, in a continuous-flow process, if the first withdrawal from the bin is started say 3 hours after the first input into the bin from the NPD and if the inflow and outflow are carefully balanced, then all the rice grains in the mass would be treated for exactly 3 hours, no matter for how many hours or days or weeks the batch is operated nonstop.

It should also be noted that the system is very efficient with very little moving parts. Once the paddy is lifted from the ground level to the top holding bin, the rest of the grain movement is entirely by gravity other than within the NPD. In the process as operated in the rice mill mentioned, the treated paddy when milled showed a milled product that looked and cooked exactly like aged raw rice. In addition, the head rice yield after milling was reduced by only about 2–2.5 percentage points compared with the same paddy milled without the treatment.

A few more points may be worth mentioning here:

(a) The dry-heat AAR process was originally conceived for processing of milled (raw) rice in mind. The idea was that if milled rice could be directly cured by dry heat, then the additional milling breakage encountered in the steam-curing process of paddy could be avoided. However, it has been subsequently verified that the process can be equally well applied to (new) paddy or even brown rice. The detailed process described above was one in which paddy was used as the raw material.

(b)   The capacity of the plant can be made as desired. NPD of any capacity along with an appropriate matching mass-flow bin and a vibro cooler can be used. The plant in the process described above consisted of three units working together in parallel each with a capacity of 5 tonnes of paddy per hour, giving a total capacity of 15 t/h.

(c)   The time of hot storage required for proper curing and the temperature of treatment are inversely related. The higher the temperature, the faster is the curing effect achieved. But a high temperature also tends to impart a faint browning effect. So the temperature is to be decided as the best compromise between the least duration of processing on the one hand and negligible browning on the other. Usually, a temperature in the range of 85°C to 90°C gives the best compromise. But this should be tested for the variety to be treated, and one should also determine the correct corresponding time of treatment. It is best to determine these conditions in the laboratory.

A small laboratory unit consisting of a preferably steam-jacketed, cylindrical, closed, rotating, steel unit of capacity 2–5 kg grain and fitted with a thermometer can be used. The sample of the paddy can be heated in it to different temperatures, and the hot grain is transferred to a preheated metal container with a tightly fitted lid. The container is now transferred to a thermostatic oven adjusted to the same temperature and left aside for different times. The sample can then be cooled, milled and cooked to check for optimum curing and the grain discolouration. The best temperature and time of treatment thus determined can then be applied to the production unit. One should also remember that the required time of hot storage is quite substantial (2–3 h) when the paddy is fresh after harvest but progressively decreases as time passes (say perhaps half hour after 8 months).

Another matter of much theoretical interest is the relation of the process to temperature. It has been mentioned above that the rate of curing or artificial ageing by the application of heat is strongly influenced by the temperature (and of course by the time of treatment as well). The higher the temperature of treatment, the faster is the rate of artificial ageing. Thus if a sample heated to 90°C requires let us say 3 hours of heat treatment to achieve a particular stage of ageing, it may need perhaps 10 hours to attain the same stage at 80°C.

One may recall that this relationship between rate of ageing and temperature (range 0°C to 40°C) was originally found to hold true for the natural process of rice ageing. Surprisingly, the same relationship is found to hold for artificial ageing as well (tested between 40°C and 100°C both in laboratory experiments and in production scale). In fact a single Arrhenius-like equation has been found to hold for the natural as well as the artificial (by dry heat) process of ageing (Fig. 6.5). This identity of a natural and an artificial process raises intriguing questions about the nature of the ageing process itself. If the same relationship holds at 20°C or 80°C or 100°C, then surely the process is not a biological (enzymic) one. Is it then a purely physical or chemical or physicochemical change? Why then a similar change is not encountered in other cereal grains?

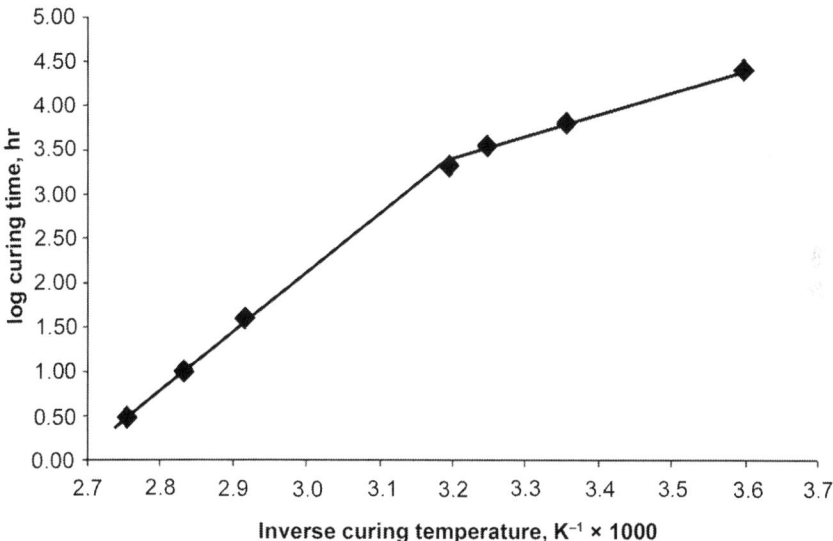

Figure 6.5 Arrhenius-like plot between ageing time and inverse temperature (Kelvin) [Reproduced, with permission, from Bhattacharya (2011)]

6.5   Preventing/retarding ageing of rice

If south Asians are rather obsessed about their dislike of new rice, the opposite is the case in east Asia. The Japanese in particular are eager to maintain the 'freshness' of new rice as long as possible. One process by which the ageing of rice can be delayed is already in operation. Other possible lines of approach too are under investigation.

The method already under practice is cold storage. It has already been discussed above that the ageing process, whether natural or artificial, is heavily dependent on the temperature. The higher the temperature the faster is the rate of ageing, and conversely the lower the temperature, the slower is the process. This finding has been taken advantage of in Japan. A general practice has evolved there wherein rice is stored under cold conditions. One may remember that Japan is a temperate country (around 40° N). The harvested rice is therefore already quite cold and it has to be stored anyway. All it therefore needs is that the storage structure be under good insulation, so that the low temperature can be maintained as long as possible. And air conditioning is maintained anyway. It has been found in practice that if rice is stored below 15°C, the rate of ageing of rice is sufficiently low, so that it remains acceptable for several months. This process is widely in practice in Japan. The process has another advantage. Insect activity is practically nonexistent below 15°C. Even fungal activity remains low under cold storage. Therefore, rice can be stored for fairly long periods without any treatment with insecticides and also with somewhat higher moisture contents (up to 15%).

If the above is a process to prevent or retard the ageing of rice, possible methods to reverse the effect of already aged rice are also under study. In the researches carried out to investigate the causes of rice ageing, the effects of various enzyme treatments were already investigated. The aim was to study the possible roles of protein, disulphide cross-linking and cell walls in the ageing process. The idea was that if an enzyme treatment reversed the effect of ageing, then surely the target process was involved in the ageing process. Thus treatment with cellulase, pectinase, endoxylanase seemed to reverse the hardening of cell walls and restore the cooking quality of original rice. Similarly, treatment with β-mercaptoethanol, dithioerythritol, cysteine or reducing agents in general (such as sodium sulphite), which could break disulphide bridges, seemed to restore the original cooking quality of rice. In another approach, it was speculated that the major change during ageing occurred in an ultra-thin layer at the surface of the milled rice. As evidence, overmilling of aged rice was claimed to restore its original cooking quality. Similarly, adding a reducing agent such as sodium sulphite to rice during its cooking was also claimed to improve the undesirable taste of old rice. One can see that possible practical ways to reverse the effects of ageing of rice lie well in these investigational results. It is not unlikely that some such method can be converted into a practical process some day in the future to reconvert aged rice into a state when it cooked like new rice.

Further reading

The subject of rice ageing has been critically reviewed, with full reference to all earlier publications, in:

• Bhattacharya (2011)

The dry-heat AAR process for accelerated ageing of rice is described in:

• Bhattacharya (2013a)

7
Parboiling of rice

7.1 Introduction

7.1.1 Two types of rice

The English word 'rice' for the seed of the plant *Oryza sativa* L. and its milled product can be a bit ambiguous. Most native languages in Asia have a specific name each for the seed (paddy, padi, rough rice), its milled product (milled rice, white rice) and cooked rice (table rice). This is not quite so in English. In fact, in English, the word rice is also used as a generic name covering all three of paddy, white rice and cooked rice. The situation may become still more complicated when it comes to identifying the type of the rice. Actually there are two types of rice, namely raw (i.e., raw not in the sense of being uncooked, but in the sense of being non-parboiled) rice or white rice and parboiled rice, for each of which there is a specific name in all Indian languages. This is in fact true of entire south Asia or the Indian subcontinent, where parboiled rice is very much a household commodity unlike in other parts of the world.

Thus rice, especially in the south-Asian context, comes in two forms – raw and parboiled. Parboiled rice is nothing but 'par'-tially 'boiled' (i.e. partially cooked) rice. In other words, it is rice precooked within the husk (i.e. in the paddy form) in a limited amount of water so as not to disturb the size and shape of the grain, and then dried back. So despite the overall similarity to the original raw rice, the grain gets a substantially different property profile.

7.1.2 Parboiled rice is typically a south Asian product

The association between south Asia and parboiled rice is a striking one. Here is an exact mirror image of the situation with rice per se. Rice is grown widely throughout the world but is heavily concentrated in the southeastern corner of Asia. Over 90% of world's rice is grown in south, southeast and east Asia. In a curious way, a parallel situation exists with respect to parboiled rice. Rice is widely parboiled throughout the world. Yet over 90% of world's parboiled rice is produced and consumed in south Asia (the Indian subcontinent).

Almost the entire amount of rice in Bangladesh, Sri Lanka and in several rice-consuming states of present-day India (Assam, Bengal, Odisha, Tamil Nadu and Kerala) is parboiled. Substantial proportions of rice in a few other rice-consuming states of India and that in Nepal are also parboiled. Overall it is considered that about 55% of rice in India and about 65% of rice in south Asia is in the parboiled form. This is shown in Fig. 7.1. A fifth of world's rice is parboiled.

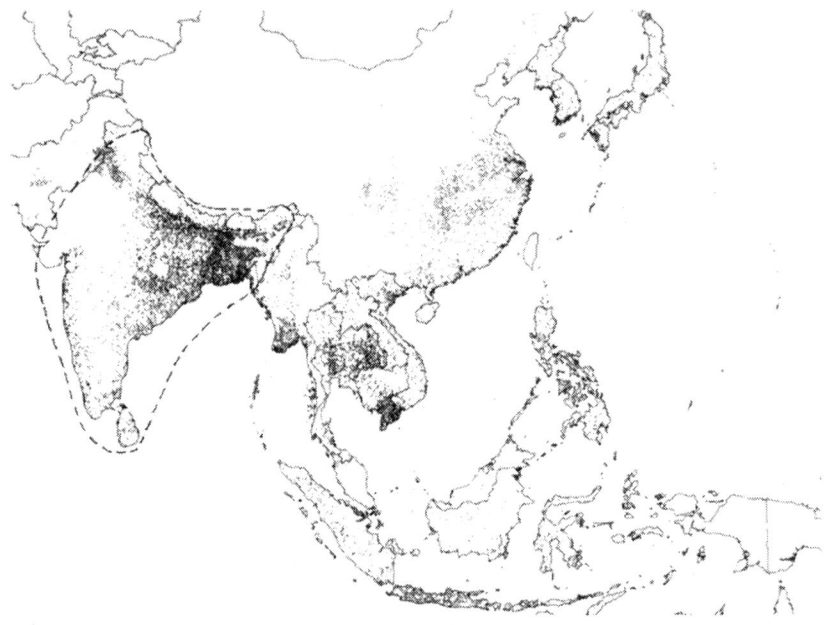

Figure 7.1 Raw and parboiled rice areas in south Asia (impressionistic). South Asia is the home of parboiled rice. Over 90% of world's parboiled rice is made and consumed here [Reproduced with permission from Bhattacharya (2011)]

Thus, the present day concentration of parboiled rice in south Asia is striking and deserves some enquiry.

7.2   History of parboiled rice

Not much is known about the history of parboiled rice. It is generally believed to have originated in ancient India. But little is known about when and how. Recently some evidence has been found about its mention in ancient Tamil (one of the Indian languages) literature in the early part of the present era. The process probably existed even earlier, but the matter needs historical investigation. An interesting feature is: the product has similarity with a similar product of wheat, namely bulgur. Bulgur is supposed to have originated around eastern Mediterranean (Middle East, northeast Africa, Turkey and southeastern Europe) in ancient times and commonly used by Hittites, ancient Greeks and is said to be still common in those regions. The similarity is obvious. Whether there is any commonality in origin or in history is again worth investigating.

7.2.1   The south Asian connection

Rice is heavily concentrated in the 'Rice Country' in south, southeast and east Asia. In this context, the heavy concentration of parboiled rice in south Asia, and yet its total absence in the remaining two regions of the Rice Country (namely southeast Asia and east Asia) is a mystery. Two possible explanations come to mind. One is the possible link with bulgur mentioned earlier in India's adjoining areas of central Asia and eastern Mediterranean, and the possible link with the often-disputed theory of Aryan migration from central Asia into India in ancient times. Do bulgur and parboiled rice have a similar origin in history? If so, that may explain why it is prevalent in south but not in southeast and East Asia.

The other possibility is related to the type of rice to which people in the three sub-regions of the 'Rice Country' are traditionally habituated. Although a distinct entity as such, the Rice Country in Asia is not entirely homogenous. As explained in Chapter 1, there are three distinct types of rice specifically associated with the three sub-regions or zones of the Rice Country. The rice in eastern Asia is low-amylose type and cooks very soft and sticky. That in Southeast Asia is largely intermediate-amylose type and cooks intermediate. But the rice in south Asia is high-amylose type and cooks firm and fluffy. It looks significant in this background that parboiled rice is so popular precisely in south Asia and there alone. It seems as if a large section of the people in south Asia were not satisfied with the type of high-amylose rice available and wanted still more firm and fluffy cooked rice. It is significant that this is precisely what parboiling also achieves. The cooked rice from it is very firm and discrete. Hypothetically speaking, parboiling thus acted as if it figuratively added more amylose to the local rice to satisfy the taste preference of a part of natives in south Asia. Parboiling thus suited the shifting trend of taste preference of rice from northeast to south Asia within the rice countries of Asia (Fig. 7.2).

Be that as it may, with this background, parboiled rice was strictly an isolated, regional product. It was produced in every home in large parts of south Asia and consumed locally by the natives but was probably totally unknown in the rest of the world. A few historical events around the turn of the 19th to 20th century brought parboiling to the notice of rest of the world.

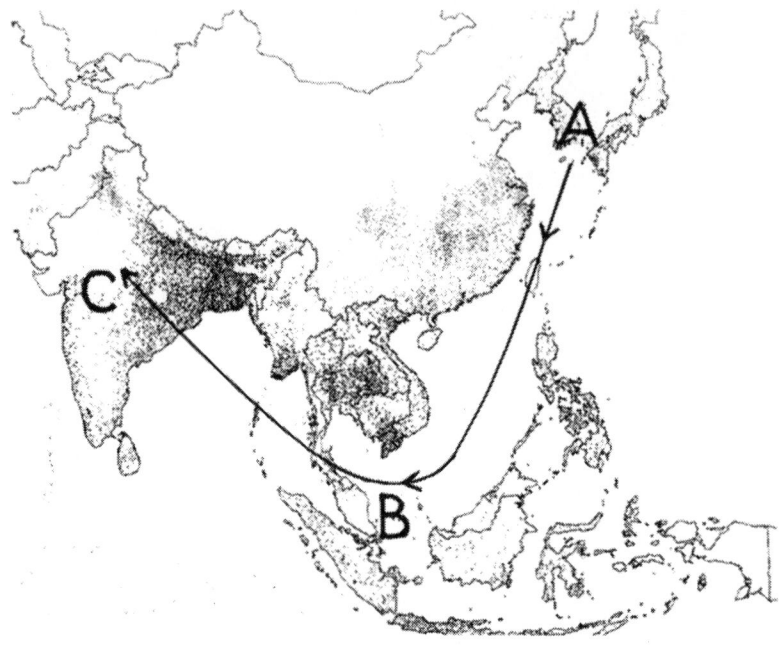

Figure 7.2 Home of the world's rice (Rice Country; impressionistic). More than 90% of world's rice is grown and consumed here. The arrow shows the direction of gradation in the sensory quality of cooked rice from soft and sticky (zone A) to firm and fluffy (zone C)

7.2.2   The beriberi–thiamine saga

One of these events was a curious but alarming medical problem, the sudden widespread appearance of the scourge of beriberi in Asia. Medical, nutritional and public health authorities suddenly noticed towards the turn of the 19th century the sudden widespread appearance of the menace of beriberi in the rice-eating areas of Southeast Asia. A series of thrilling epidemiological studies over two to three decades ultimately led to the finding that the disease was caused by a deficiency of thiamine (vitamin B1) in milled rice. What actually happened was that machines for milling of rice became available around that time. Use of these machines led to excessive removal of bran to make rice more white. This excessive milling led to loss of the vitamin from rice which caused the deficiency disease, beriberi. The vitamin was mostly located in the outer layers of the grain, so was easily removed by overmilling by machines.

What was still more curious was the fact that wherever parboiled rice was eaten, there was no beriberi. The story of this trail of investigation reads almost like a detective story. One famous example may be cited here as illustration. A group of Indian road labourers in Malaya was divided into two parts. One part was given raw rice, the other part parboiled rice. Within six months, the raw-rice part started getting beriberi but not the other group. Then the rice was reversed, the raw group being now given parboiled rice and the parboiled group raw rice. The beriberi in the earlier raw group soon disappeared but started appearing in the earlier parboiled group. There are many such thrilling accounts of beautiful epidemiological research which is worth reading. Anyway it finally led to the finding that the vitamin B1 got firmly attached to the rice kernel when it was parboiled and was therefore not removed easily by milling. That is how it prevented beriberi. Parboiled rice was thus nutritionally superior to raw rice.

7.2.3   Emergence in the world stage

With this finding, parboiled rice suddenly came to the attention of scientists and rice industries outside south Asia. Many scientists started research investigations on it and rice companies started looking into it. Then they noticed other technological benefits of parboiled rice. One amazing beneficial property was that the breakage of rice during milling was dramatically reduced if it was parboiled. Again, as parboiled rice gave harder cooked rice, cooking of rice was also easier for people not used to rice. Parboiled rice would not easily overcook, so timing of its cooking was not so critical. Another interesting property was its suitability for canned rice. Western industries at that time were trying to produce canned cooked rice. But the matter was difficult because the rice got too much mashed. It was now found that if parboiled rice was used instead of raw rice for the purpose, it gave a better canned rice product. All these beneficial properties made parboiled rice a darling of the rice industries in Europe and America and scientists there began intensive research in parboiling of rice. That is how parboiled rice suddenly became more widely known throughout the world as a material of much scientific and industrial interest in the early decades of the 20th century.

This wide scientific interest had its echo in India as well. Fledgling scientific institutions were then emerging in India especially in Kolkata (then Calcutta) and Bangalore. In the background of the extraordinary qualities of parboiled rice, interest in it as a subject of study arose here as well. Apart from interest in its increased vitamin B1 and reduced milling breakage, the main interest here was in another matter, viz. the production process. This was partly prompted by international interest. The international agencies strongly advocated popularising of parboiled rice as an antidote to beriberi. In that context, a strong interest arose to modernise the primitive, somewhat unhygienic parboiling industry, and this was repeatedly raised by the international agencies. This work was finally taken up over a long period of time in India in the centres mentioned and finally solved in the Central Food Technological Research Institute (CFTRI) in Mysore towards the end of 1950s. The long saga of these two lines of scientific research is available in some treatises and is of much interest from the historical point of evolution of scientific research, in India as also elsewhere.

7.3   Production of parboiled rice: Processing conditions

It has been mentioned above that parboiled rice is nothing but a kind of precooked rice. It is rice that has been cooked within the husk, i.e. in the paddy form. As the intended product is also a kind of rice, the cooking is done in a limited amount of water, so that the size and shape of the grain are not much disturbed. To this end, cooking is separated into its two parts. Cooking normally involves simultaneous application of water and heat. But in the case of parboiling, the two applications are kept separate. The paddy is made to hydrate (i.e. allowed to absorb water) first, and then heat is applied to the soaked grain to complete the cooking. The process of parboiling thus involves three steps: soaking (absorbing water), heating (to complete the cooking) and drying (to dry back the paddy). The necessary conditions for these three steps are described below. The basic principles and the theory of the different steps are discussed here. The actual production processes are described in the next section.

7.3.1   Soaking

The first step in the process of parboiling is to soak the paddy in water, so that it gets adequately hydrated. About 90% of the dry matter of rice is made of starch. Therefore, starch properties play a crucial role in the behaviour of rice. One important property of starch is its gelatinisation temperature (GT). This is the temperature at which starch gets cooked (becomes swollen, pasty, slimy) when heated with water. Generally, the GT varies between 60° and 75° C in rice. The majority of Indian rice varieties have a GT of around 70°C to 74°C.

The temperature has a strong effect on the speed of hydration. The rate of hydration (i.e. water absorption) of paddy in water, like most other reactions, increases with increasing temperature. However, gelatinisation plays a crucial role here. This can be seen in Fig. 7.3. At temperatures below the GT (about 70°C), rice goes on absorbing water at an increasing rate with increasing temperature, but gradually decreasing with time of soaking. Ultimately, it reaches an equilibrium around 30% moisture (wet basis). But at temperature greater than GT, the pattern changes. After an initial period, the rate of hydration now increases and goes on increasing at accelerated rate. This is because at such temperatures, the starch in rice starts to get gelatinised, whereby its hydration goes on rapidly increasing. One more fact is to be noted. Whenever the moisture content of paddy exceeds about 30–32% moisture, the husk bursts leading to possible deshaping of the kernel. As can be surmised from the discussion above and Fig. 7.3, this can happen only if paddy is soaked at a temperature above the GT.

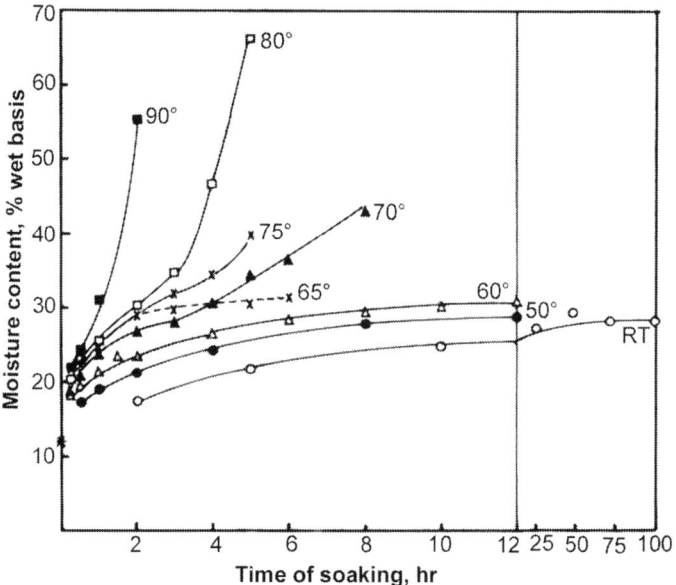

Figure 7.3 Hydration of paddy at different temperatures. RT = room temperature [Reproduced with permission from Bhattacharya and Subba Rao (1966 a); ©American Chemical Society]

Now for the purpose of parboiling, the intention is to allow the paddy to get uniformly saturated with water, i.e. to have around 30% moisture. It is clear from the discussion above and from Fig. 7.3 that any temperature below the GT would be suitable for the purpose from the standpoint of avoiding oversoaking and grain bursting. Timing too would not be critical. But at too low temperature, time required to achieve saturation would be long. So, with food (paddy) remaining mixed with water for long, and at a fairly low temperature, there is bound to be microbial growth and fermentation leading to production of bad odour. So temperatures below 60° to 70° may be suitable for hydration but not desirable in practice at all. At temperatures above the GT, on the other hand, soaking would no doubt be fast. But there would be every danger of oversoaking. The outer layers of the paddy grain would get cooked and would go on absorbing moisture. So the paddy would get oversoaked and burst and deshaped even before the centre of the rice was properly hydrated. Therefore, soaking at temperatures above GT too is not at all suitable for parboiling either. It can be done, but then the timing would be critical. The water must be drained out as soon as the paddy attained an aggregate moisture content of around 30%, and then it would have to be tempered (i.e. allowed to rest) to allow the moisture to get equalised within the grain. The ideal temperature for soaking for parboiling therefore is just around or below the GT (generally around 70°C for Indian paddy varieties).

There are certain varieties of rice that have certain attractive qualities but have a low GT and hence pose problems in soaking. For example, Pusa basmati 1 is a well-known basmati variety which yields parboiled rice of excellent quality but its GT is a little less than 60°C. It therefore requires soaking at 55°C to 60°C and needs about 10–12 hours of time to achieve saturation. This prolonged soaking at such low temperatures is very conducive to bacterial anaerobic fermentation and consequent off-odour. It has been found that bubbling air into the tank during soaking helps in avoiding the fermentation. Fermentation is totally absent whenever the soaking is done at 70°C or more. At temperature below that, it is always desirable to use bubbling of air in addition to maintaining strict hygiene.

The rate of hydration of the paddy can be influenced to some extent by application of vacuum and hydrostatic pressure. There is some air entrapped in the hairy surface of the paddy. This air partly prevents the immediate wetting of the paddy surface in the water. So there is some delay in the absorption of water by the paddy. If a vacuum is applied to the paddy before running in water, then that entrapped air is removed and the paddy is immediately wetted. Applying vacuum therefore hastens the hydration process to some extent. It may be noted here that passing steam in paddy for pre-steaming (see below) achieves the same purpose. The soaking time is shortened if paddy is pre-steamed before soaking. Similarly, if air pressure is applied in the soaking vessel above the water level, then the water becomes pressurised and so penetrates the paddy grain faster. Application of vacuum or pre-steaming or pneumatic pressure (i.e. air pressure) thus helps in the soaking process by reducing the time of hydration.

7.3.2   Steaming

Once the paddy has been adequately hydrated, it has now to be cooked. For this the excess water is to be drained out first (otherwise it will become like cooked rice). Since the paddy is already hydrated to saturation, cooking of the starch in the grain now only requires application of heat. Heating the soaked paddy by any manner to a temperature of 80°C or more should be able to perform the cooking. However, in practice, the simplest and easiest way of heating is to pass steam into the soaked paddy. A boiler is the easiest method to generate heat in one place (by burning a fuel) and carry it in the form of steam to wherever one wants the heat. It is only where a boiler is not available, that other methods of heating such as by contact heating (conduction), microwave heating, heated air have to be considered. A minimum amount of steaming to raise the temperature to 100°C is enough for the cooking or gelatinisation. However, the severity of steaming or heating process has an effect on the product quality. The more severe the heating or steaming (high steam pressure or longer time), the greater the degree of gelatinisation but also of discolouration. More severe cooking makes the parboiled rice progressively harder. This hardening helps in reducing milling breakage but increases the time required to cook the rice at home. At the same time, one must remember that a faint yellow-brown discolouration is produced during parboiling. This is normal. But this discolouration too increases with increasing degree of steaming or heating. So the time and pressure of steaming have to be adjusted according to the product quality desired.

7.3.3   Drying

Paddy has approximately 30–32% internal moisture (wb) within the grain at the end of soaking. Another approximately 5% water is present as external water adhering to the grain. Also, when steam is applied to the wet paddy, initially there is some condensation and that also adds to the moisture. The parboiled paddy at the end of steaming therefore has a moisture content of about 35–40%. It has to be dried immediately to a safe storage level of approximately 13%.

The basic principles of drying of parboiled paddy are the same as for drying of field paddy as discussed in detail in Chapter 3. But a few points should be specifically noted. One, the moisture content of the paddy is so high that any option of slow air drying is entirely ruled out. Drying has to be done either in the sun or by heated air in a continuous-flow dryer. In a tropical country like India (one should not forget that 90% of world's parboiled rice is produced in the Indian subcontinent), with parboiled rice produced in thousands of small rice mills with a daily capacity of perhaps 50–200 tonnes of paddy and with plenty of relatively inexpensive labour available, sun drying is very much a viable option. Nor is it as inefficient as some ivory-tower experts instinctively assume. Sun drying of parboiled paddy on large masonry floors was the common practice in thousands of rice mills in India until approximately the turn of the millennium (Fig. 7.4). However, use of LSU dryers has been gradually making inroads since then.

Figure 7.4 Sun drying of parboiled paddy in a rice mill. Top – paddy being arranged into furrows. Bottom – paddy heaped and covered at night or during rain or for rest [Reproduced from Bhattacharya and Ali (1970)]

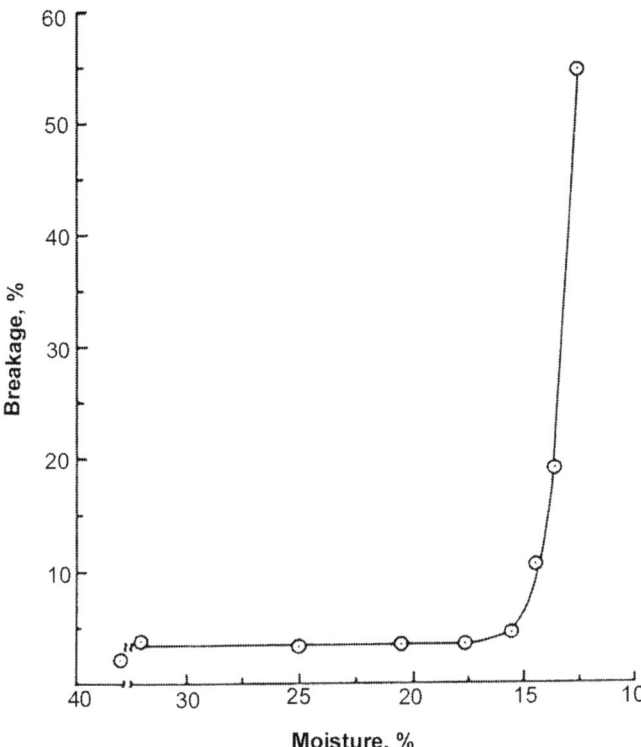

Figure 7.5 Effect of drying on milling quality of parboiled paddy. Paddy dried with air at 82°C in an LSU dryer to various moisture contents, then finish-dried in the shade and milled [Reproduced with permission from Bhattacharya et al. (1971)]

The point of departure for the method of drying parboiled paddy from the systems of drying field paddy is in the number of passes during heated-air (or sun) drying. In drying of field paddy, whenever moisture content is more than 18–20%, drying must preferably be performed in several passes (more than two). But in the case of parboiled paddy, in view of uniformity of moisture content in all the grains of the batch, two passes are sufficient in theory. It has been repeatedly verified that no damage occurs to the grain upon rapid drying of parboiled paddy until the moisture content drops to about 16% (Fig. 7.5). This is true no matter how fast this first stage of drying is done. Damage then occurs at an accelerated rate and milling breakage may go up to very heavy levels (50–100%) if the process is carelessly operated. So the paddy must be adequately tempered at approximately 16% moisture level. About 4 hours of tempering is sufficient if the paddy is somewhat hot, otherwise 8–12 hours may be needed. If properly tempered at this stage, the paddy may then be further dried down to approximately 13% moisture without appreciable additional damage. However, if there is already adequate arrangement for additional tempering and drying in several stages, as in certain ultramodern systems, there is no problem at all. But drying in two stages with tempering at around 16% moisture is a minimum requirement. This has been well tested and demonstrated in production level both for sun drying and heated-air drying.

Two additional points should be noted with respect to practical drying procedure for parboiled paddy. First is in relation to drying in LSU dryers with heated air. It has been mentioned that no damage occurs to the rice even with drying at a very high temperature (air temperature of 100°C or more) until the paddy moisture drops to about 16%. This fact provides an opportunity for adopting a relatively simple drying procedure. In normal drying installations in West, it is usual to have a series of dryers (two, three or more) with tempering bins between each pair operating in a continuous system. While this is a proven and efficient system, a simpler and less elaborate system can be used equally satisfactorily for parboiled paddy drying. Thus, one can satisfactorily dry parboiled paddy with a single LSU dryer and without any tempering bin in a batch system in rice mills having a modest capacity. One can steam the soaked paddy (preferably in a continuous system) and allow the wet paddy to fill up the dryer. Once filled, either the steaming can be stopped or the steamed paddy can be diverted to an alternate dryer. The heated air in this first dryer then can be started and the paddy is allowed to continuously recirculate within the dryer until the moisture content drops to approximately 16% (which would take about 4 hours). Drying can then be stopped and the paddy is allowed to temper within the dryer itself for 4 hours (instead of taking out the paddy into a tempering bin and later refilling after tempering, which would in any case consume 1–2 hours). The paddy can now be dried for the second stage within the same dryer at a relatively lesser temperature (at about 70°C, requiring about 1 hour). This batch system is quite efficient and cost effective for installations with modest capital density.

The second point with respect to drying procedure of parboiled paddy relates to the handling of the paddy just after steaming. The paddy at this stage is soft with the husk slightly split. Any handling of this paddy is therefore liable to damage the shape of rice and also cause contamination of equipments and create unhygienic surroundings. It is therefore preferable that the paddy after steaming (post-soaking) is fed by gravity directly into a quick dryer. This dryer would remove the surface moisture and close the split husk and make the paddy free-flowing. A rotary dryer or a vibro-fluidised dryer is ideal for this purpose. This additional dryer would no doubt add to the cost to some extent. But it would improve the efficiency of the production system. The main purpose of this dryer is not so much to add to the dryer capacity but to make the paddy amenable for handling. Removal of 5–10% moisture is sufficient to make the paddy free-flowing and suitable for handling. This partially dried paddy can then be taken to an LSU dryer for the main drying.

7.3.4   Milling

Parboiling is complete with drying of the processed paddy. Milling is the next step in production of parboiled rice. But that is similar to the usual process of milling. However, there are some minor differences. One, the husk normally gets marginally opened after steaming of the parboiled paddy. The dehusking of the paddy is therefore easier after parboiling. Two, but further polishing is a little difficult. The bran of parboiled rice is a little more oily than normal. Therefore, the screen mesh of the polisher has a tendency to get clogged during whitening of parboiled rice. This aspect has to be taken care of either by maintaining a greater pressure in the whitening unit, or by increasing the mesh openings of the screen slightly. Millers also often add a little chalk or husk to the dehusked rice during whitening. In any case, whitening of parboiled rice requires more energy than whitening of raw rice.

7.4   Production of parboiled rice: Production methods

The basic principles of the different processing steps have been explained in the above section (Section 7.3). It is now possible to describe the actual production processes in some detail. However, the process details should not be considered in isolation. These should preferably be visualised in conjunction with the basic principles of the steps enunciated above.

Parboiled rice is an ancient product. And now a very sophisticated modern industry has also come up on it. So parboiled rice has been made by very primitive to highly modern systems. Besides, as mentioned before, rice can be parboiled in various ways. There are three basic systems of parboiling:
  (a) Soak-drain-steam-dry process,
  (b) Pressure parboiling (low-moisture parboiling) and
  (c) Dry-heat parboiling (soak-drain-HTST heat-dry process)

7.4.1   Home- and cottage-scale methods

Initially, parboiled rice was produced at homes for centuries and the practice continued in villages in south Asia almost until sometime in the second half of the twentieth century. As there was no question of steam generation at homes, the entire process was based on hot water. Endless variations existed. Perhaps the paddy was boiled in water till the husk split; perhaps it was soaked in water and then boiled; perhaps soaked in warm water, drained and then steamed from a little water at the bottom of the vessel; and finally dried in the sun. These were quite unsophisticated products and either had prominent 'white belly' (uncooked white centre) or were burst and deshaped. However, it must be remembered that these products were hygienic and without any microbial contamination.

Apart from home processes, some primitive rudimentary scale-ups also existed. For instance, paddy might be taken in a big box-like iron container with a false bottom in between. Initially water would be taken up to above the paddy level and heated from below to soak the paddy. After completion of soaking, water would be drained out to below the false bottom and now heated again to steam the soaked paddy. Finally, the paddy would be sun dried. Sometimes some primitive-level, home-made boilers also would have been used. All these products, meant essentially for consumption at or around the homes, must be considered of poor quality by our standards but were hygienic and wholesome.

7.4.2   Soak-drain-steam-dry processes

This is the fundamental system of parboiling. The lion's share of the world's parboiling processes belongs to this category. Here paddy is first soaked in water to hydrate to saturation. The water is then drained out. Then steam is passed into the soaked paddy. The paddy so parboiled is finally dried. While this is the basic process, this can be carried out in various ways and at various levels as described below.

Single-boiling process

With the development of towns, cities and market towards the end of the nineteenth century, large-scale production of parboiled rice for the market became necessary. Millers took the simplest way out to build huge ground-level masonry tanks or pits (capacity 5–20 tonnes of paddy) in which paddy was soaked in cold ambient water. Because of the low temperature, it took about 3 days for the hydration to be completed. This prolonged soaking in ambient water invariably caused microbial fermentation and production of bad smell. The smell did not go even after the paddy was steamed and dried. The soaked paddy was taken out of the pit in baskets. It was now steamed in small hopper-bottom kettles (capacity 100–250 kg paddy) provided with an upright steam pipe open at the bottom. The steamed paddy was then dried on the yard.

This process yielded a product which was good as far as parboiling per se was concerned. But it was rather unhygienic and invariably smelling. The process continued in the Indian subcontinent at least for half a century or more (until at least the middle of the twentieth century). But it seems by now to have been nearly abandoned. The mills might be producing up to 500 tonnes of parboiled paddy per day in season. This process was called 'single boiling' because steaming was done only once. A traditional parboiling installation in a south Indian rice mill is shown in Fig. 7.6.

Figure 7.6 Traditional parboiling (single boiling), showing masonry soaking tank (foreground) and steaming kettles (background right) [Reproduced with permission from Bhattacharya (1985)]

Double-boiling process

Soaking in cold water as above require 3 days. Some enterprising millers found a way to reduce the soaking time. In this process, paddy was steamed first in the steaming kettles before being dumped into the soaking pit. Cold water was run into the pit once it was full. The water got somewhat heated upon coming in contact with hot paddy, thereby reducing the soaking time to 20–40 hours. The rest of the process of steaming and drying remained the same. Because the process involved steaming twice (once before soaking and again afterwards), it was called 'double boiling'. The first steaming of the paddy was merely an indirect way to heat the water.

Although ingenious, most of the problems of the original process remained here as well. Although the water got heated, the temperature was not sufficiently high nor uniform to prevent microbial fermentation and the smell problem remained. Besides, water in the big soaking pits became stratified by temperature; hot water came up and cold water went down. So, heavy fermentation occurred at the bottom. Despite these drawbacks, this has been the process most widely followed in India until towards the end of twentieth century.

Modern hot-soaking processes: CFTRI process

Once parboiled rice was introduced into the world stage through the episode of beriberi and vitamin B1, many rice companies in the West became interested in this wonder product. They took up internal research and developed their own modern processes. But these processes were patented and remained closed to the outside world. In the meantime, strong suggestions came from international agencies to popularise parboiled rice in view of its improved nutritional and anti-beriberi status. For this, it was essential to eliminate the fermented bad smell and to make the product hygienic. A good deal of research was therefore undertaken with this object especially in India over a few decades in the 1930s to 1950s. Finally, the solution emerged that soaking should be carried out at a high temperature (near 70°C). In that case, the soaking could be completed in 3–4 hours and the undesirable smell was totally eliminated. However, soaking at a high temperature could not be accomplished in the traditional set-ups and required some changes in the equipments and technology.

The first such process was developed in the CFTRI in Mysore and released by the Institute for public use by the end of 1950s. In the process as developed in CFTRI, soaking is done in several 2–4 tonne hopper-bottom, above-ground, cylindrical, mild-steel tanks. Each tank is provided with a steaming manifold made of a central and several branch pipes, the pipes being closed at the ends but carrying perforations for steaming (Fig. 7.7). Water is first taken in the tank and steam is passed to heat the water to a little above 90°C. Or hot water is run in from a hot-water tank. Paddy is then dropped into the water and the water is recirculated within the tank to equalise the temperatures at different layers. The final water temperature after mixing with paddy comes to about 70°C to 75°C and the paddy is allowed to soak for 3–4 hours as appropriate with intermittent recirculation. Once the soaking is completed, water temperature by now would drop to 63–67°C and danger of over soaking is avoided. Alternatively, water temperature is maintained at 70°C by intermittent injection of steam (outside the soaking tank). In that case over soaking has to be avoided by strict timing. The water is now drained out. Steam is now passed through the pipes until steam emerges both from top and bottom of the tank. Parboiling is now complete and the paddy is taken out and sent for drying (either on to the yard for sun drying or to the LSU dryer).

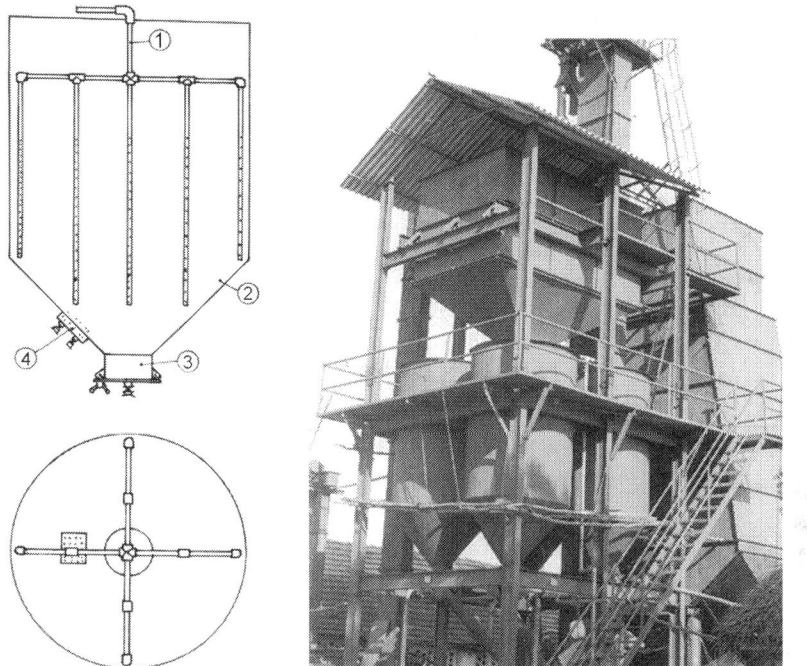

Figure 7.7 Parboiling tank of CFTRI process. Left – Schematic drawing; Right – Tank at an installation [*Courtesy:* Sri Valli Engineering Works, Tiruvallur, Tamil Nadu]

It may be noted that this essentially simple and unsophisticated, yet satisfactory, technology was developed to suit the status of the Indian industry (which was and still is by far the major arena of parboiling in the world). It is certainly not a very sophisticated process in the context of a highly modern industry. But it gave a hygienic and totally smell-free product and suited the state of the Indian industry.

Modernised double-boiling process

Strenuous efforts were made from 1960s onward by the Indian government and other agencies to popularise the improved CFTRI hot-soaking process. But somehow, due to reasons of their own, this process did not become popular among rice millers. One possible reason could be that steaming of a large batch of paddy (3–5 tonnes paddy in a tank) requires setting up of a rather large-capacity boiler, which probably millers were reluctant to install. This reluctance may originate either due to problems of maintenance or due to stringent government regulations. On the other hand, steaming in small kettles (of up to 500 kg paddy) calls for only a relatively low-pressure boiler, for which regulations were less stringent.

Figure 7.8 Three-tier 'modernised double-boiling' system of parboiling
[*Courtesy:* Indus Food Products and Equipments Ltd., Kolkata]

However, an intermediate level of improvement has been carried out by using a somewhat modernised version of the old 'double-boiling' process. The process is essentially the same as the old one. But instead of soaking in a huge pit, very conducive for microbial fermentation, a better system has evolved. In this, the pre-soak steaming, the soaking and the post-soak steaming equipments are arranged in three vertical tiers. The pre-steaming is done in small hopper-bottom kettles on the top. Soaking is then done in a bigger but not a huge tank at the second tier. Finally, the second steaming is done at the lowest tier again in small steaming kettles. The system has gone through several stages of improvement. The soaking tank has become progressively smaller and finally entirely closed with recirculation facility, for better control of a uniform soak temperature (see Fig. 7.8 for a latest version). The reduction in the unit capacity has been compensated by increasing the number of units. Because of the soaking being done in relatively small but closed tanks, temperature can be maintained at a fairly high level (around 65°C). If recirculation arrangements are provided and proper hygienic conditions are maintained, then bad odour can be avoided. However, proper cleaning and recirculation are not always done, so the problem may persist nonetheless. We may call this a 'modernised double- boiling system', and it has become fairly popular in India now. Some, but not all, give very good results.

Modern sophisticated methods

The essence of modernising the traditional parboiling industry is to shift from cold- or warm-water soaking to hot-water soaking. The associated equipments and technologies are a matter of choice and are not necessarily an essential part of the basic process. Several patented processes were developed in the West from the time of Second World War onwards. The industry in Europe and USA then went through experimenting with a number of technologies to achieve this goal over years and decades. Details of these processes were closely guarded secrets initially. Whatever could be known later can be found in several text books. But a majority of these processes (Avorio, Crystallo, Fernandez, Arlesienne processes) have now disappeared and one or at best two now survive. One, probably on the way out, is the so-called 'Gariboldi process' from Italy and the other is the so-called 'American process' from the USA.

In the Gariboldi process, a single rotating drum is said to be used for the soaking, steaming as well as the initial drying. The soaking and steaming are done in the drum by applying water and steam pressure, respectively. Initial drying is done in the same drum by applying vacuum. The partially dried paddy is finally dried in usual dryers.

In the American process, soaking is carried out in a series of upright, cylindrical, hopper-bottom tanks (capacity 10–30 tonnes paddy each) at around 70°C. The timings are so staggered that each tank is ready to discharge its soaked paddy exactly by the time the steaming of the previous tank's soaked paddy has been just completed. The process thus goes on in a continuous stream. Water is drained out after soaking and the soaked paddy is discharged on a conveyor belt. The belt or a pneumatic system carries the soaked paddy to the top of the continuous steaming unit. The latter is a tall upright stainless-steel, cylindrical vessel with a steep hopper at the bottom.

Inlet steam with appropriate entry and outlet valves ensure that the paddy is adequately steamed. It is then taken out and dried continuously in a series of LSU-type dryers with intermediate tempering between successive stages (see Fig. 7.9).

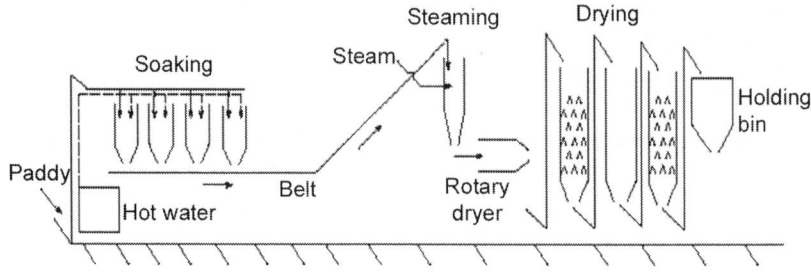

Figure 7.9 Modern 'American system' of parboiling (impressionistic)

Care is taken in these modern processes that the soft post-soaking or post-steaming paddy is handled in such a way as not to damage the delicate grain nor to contaminate the surroundings. The soaked paddy is directly delivered to the top of the continuous steaming unit by either a belt or a pneumatic conveying system. After steaming, the steamed paddy is delivered by gravity directly into either a rotary dryer or a vibro-fluidised dryer. The surface moisture of the grain is quickly removed in this dryer, so that the paddy is no longer damaged by other conveyor systems (bucket elevators). From here onwards, the paddy is dried in a series of LSU dryers with appropriate tempering in between.

Continuous steaming unit

The few highly sophisticated modern processes prevalent in the West are of course entirely continuous. The processes in operation elsewhere, including the improved CFTRI hot-soaking system, are all batch processes. For a genuinely efficient process, continuous steaming of the soaked paddy is desirable. The continuous steaming units currently in operation in the modern automatic production systems of the West are highly sophisticated and expensive systems, including automatic valves and controls. These are therefore not accessible to relatively smaller processors.

A simple steaming unit has recently arrived which can be accessible to producers with modest means as well. In this unit, steam is passed directly into the paddy as it flows down a tall column made of a series of truncated cylindrical containers (Fig. 7.10). Steam is passed into the paddy through a simple perforated manifold (spurger system) so arranged as to expose all grains more or less equally to steam. Alternate containers are for steaming and the other for tempering. A special feature of this manifold is that the perforations in the horizontal branch pipes are arranged in such a way that the number of perforations at any point is proportional to the square of its distance from the central supply pipe (Fig. 7.11). This arrangement is specially designed to suit the geometry. In a circular container, the volume of paddy to be covered by a given perforation increases in proportion to the square or cube of its horizontal distance from the centre of the vessel. Therefore, the number of the supply holes should also increase in proportion to the distance from the centre, which helps in ensuring uniform steaming of the entire mass.

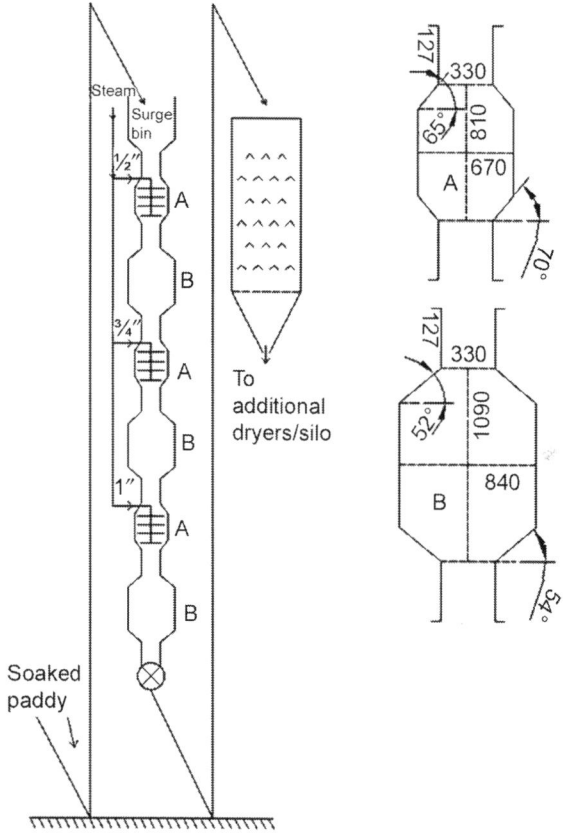

Figure 7.10 System for continuous steaming of soaked paddy (left). Right – Approximate dimensions of A and B, i.e. Bins (not to scale) [Reproduced with permission from Bhattacharya (2013b)]

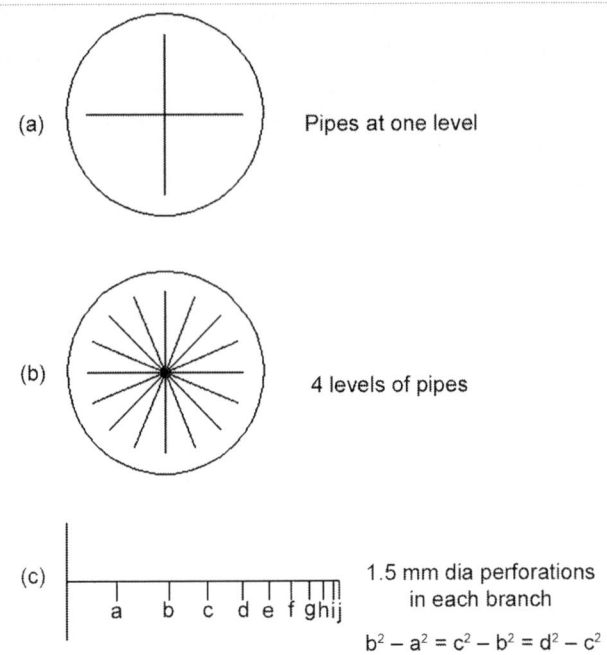

Figure 7.11 Steam sparger system in A bins of Fig. 7.10. (a) Pipes at one level; (b) Four levels of pipes; (c) Perforations in each pipe branch [Reproduced with permission from Bhattacharya (2013b)]

Another feature of the steaming unit is that the system is self-regulating. Once the rate of flow is set by the rotary valve at the top, the rest of the passage is automatically adjusted by gravity. No valve or automatic control system is required. However, adequate steam pressure in the supply line must be ensured.

7.4.3   Pre-steaming of paddy before soaking for uniform ageing control

In the 'double boiling' process, paddy is first steamed and then soaked. This pre-steaming is only an indirect method of heating the soak water. Initially it had no other objective. However, later on it was generally observed by millers that the pre-steaming, if followed by hold-up of the steamed paddy for some time, could be of help to overcome the poor cooking quality when 'new' paddy had to be parboiled. Recently, a process of pre-steaming has been developed precisely and entirely for this purpose.

Variation in product quality due to variation in raw material caused by progressive ageing of paddy with time is a perennial problem in the rice industry. The cooking–eating property of the rice continuously changes with time, and it is tough to maintain product uniformity at different times of the year. Besides, soon after harvest the crop does not yield a product of desired quality. The concept of pre-steaming of paddy before soaking has been introduced to take care of this problem (see discussion under 'steam curing' process in Chapter 6).

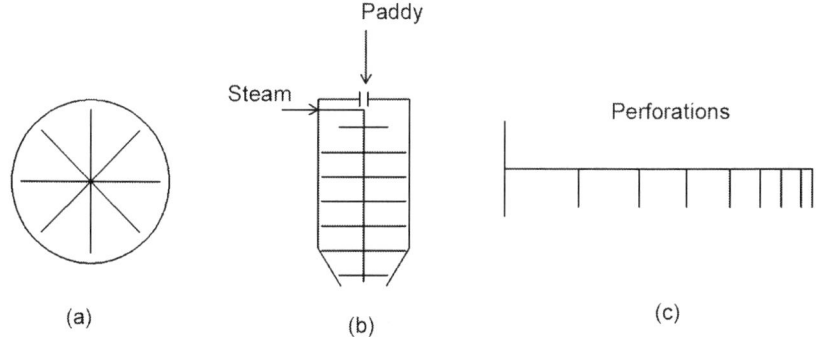

Figure 7.12 Steam sparger system for pre-steaming of paddy before soaking. (a) Pipe array at one level; (b) Steaming tank with multiple levels of pipes; (c) Perforations in each pipe branch [Reproduced with permission from Bhattacharya (2013b)]

In this process, instead of taking the paddy directly for soaking in hot water, it is first steamed and tempered (held hot). The tempering time is adjusted according to the age of the paddy. If paddy immediately after harvest requires about 2 hours of hot storage, it may need only about 20 minutes after say 8 months. But the time of hot storage should be adjusted not only for the desired cooking-eating quality, but also to ensure that there is not too much of discolouration. Pre-steaming not only changes the cooking–eating property, but also imparts a light yellow-brown discolouration. The time of hot storage should be adjusted accordingly. The steaming manifold (sparger unit) within the pre-steaming unit is arranged on the same basic principle as in the continuous steaming unit mentioned above (Fig. 7.11) to ensure that all the grains in the mass are uniformly and quickly steamed. This is ensured by arranging the manifold with adequate number of branches and appropriate number of perforations. The perforations here too are arranged according to the principle of the number of perforations being proportional to the square of the distance from the central supply pipe (Fig. 7.12). Another important requirement is that the boiler capacity has to be sufficiently high to ensure that the steaming is completed in a reasonable time (10–20 minutes). The boiler steam pressure requirement would obviously increase as the capacity of the pre-steaming unit increases. Similarly the cross-section of the main supply line of steam must be equal to or more than the combined perforation areas of the branches it serves.

7.4.4.   Pressure parboiling

A new system of parboiling was proposed towards the end of 1970s at the Paddy Processing Research Centre at Thiruvarur, Tamil Nadu, India. They found that it was not essential that the paddy be fully soaked for it to be parboiled. If paddy was only partially soaked or even merely wetted, it could still be parboiled by steaming under elevated pressure. They called this process 'pressure parboiling'. This was unfortunately a misnomer, for steaming under pressure could be and was indeed being done even under the usual soak-drain-steam-dry systems. Its special feature is actually the low moisture of the paddy during parboiling. Hence, it could well be properly called 'low-moisture parboiling'.

As the soaking was avoided in this process, it had two advantages. First, the process was fast. Second, drying was quicker and consumed less energy and cost. But it had two unfortunate defects: high discolouration and very hard rice, difficult to cook. It even attracted the sobriquet 'iron rice'. The process flourished for about a decade or so in the Punjab region of India and then was abandoned after people were reluctant to accept the rice for its hard-cooking quality.

However, other studies showed that the pressure-parboiled rice could be made acceptable by adopting appropriate conditions. Instead of using usual high-amylose rice as common in India, it was necessary to use low- or intermediate-amylose rice for the purpose. Then the cooking quality would be reasonable. In addition, it was also shown that if the processing conditions were properly adjusted to yield a rice with a white belly, then also the cooking quality remained acceptable. One great advantage of pressure parboiling was that the resultant rice was much better suited to producing canned rice than the usual parboiled rice. So ways to revive the process were welcome.

7.4.5   Dry-heat parboiling

In this system, the soaked paddy, instead of being steamed, is subjected to high-temperature, short-time (HTST) heating, usually by conduction. Superheated steam or very high-temperature air can also be used for the heating. However, conduction heating is the standard method of the HTST heating. The archetypal method is to roast the soaked paddy along with, or without, fine sand in an iron pan with vigorous stirring over an open fire. Heat is conducted from the fire to the pan to the sand and finally to the soaked paddy. The intense dry heat not only quickly gelatinises the starch and completes the cooking, but it also partially dries the grain down to about 20–25% moisture. This quick removal of excess moisture freezes the starch molecules in their gelatinised status and thus imparts certain special properties to the rice. Thus, for instance, the resultant rice absorbs quite a good amount of water easily even when put in cold water.

Surprisingly, this dry-heat parboiling method has been used in various pockets of the world for a long time without being known to others or to the outside world, including scientists doing research on rice parboiling. For example, it was being regularly used in the hill districts of Uttar Pradesh state of India for producing a kind of parboiled basmati rice for quite some time. Desiring to have hardened basmati which would not break excessively during milling and also could produce good *pilaf*, a local small-scale process indigenously sprung up, called the 'sella process'. Paddy was soaked in hot water in small cement cisterns and the soaked paddy was roasted with stirring over an iron pan heated from below, thus achieving parboiling by conduction heating. In essence this process was a small-scale multi-unit parboiling of paddy without the necessity of installing a boiler. The process was abandoned after the need for large-scale production arose with the development of international trade in parboiled basmati rice. This small-scale process then could not cope with the volume of the demand and the standard soak-drain-stream-dry process was started in Punjab and Haryana states.

A very similar process ('Chatti process') was started in Pakistan's Punjab province after the partition of India in the 1950s when a necessity arose for supplying parboiled rice to the then East Pakistan (now Bangladesh) where parboiled rice is the staple. This was a flourishing business but disappeared after separation of Bangladesh. Here again the essence was avoiding the installation of a boiler and achieving parboiling by conduction heating. In the process, paddy and water were taken in baked mud pots (*Chatties*), which were placed on smouldering burning husk beds overnight. This completed the soaking. The soaked paddy was now roasted with stirring in huge iron pans over an open fire. This dry-heat parboiled paddy was now sun-dried (Fig. 7.13) and finally milled.

Another such instance is the so-called stove process being operated in Brazil for quite some time. It is said this stove process was accidentally discovered there while trying to dry moist paddy. Moist paddy was being passed through a channel and heated by external heat for drying it. It was surprisingly found that sometimes the paddy became parboiled and the process was developed as a method of parboiling. Since then it became a standard method of parboiling being widely used there, unknown to the world.

Figure 7.13 'Chatti' process of parboiling. Empty *chatties* (1), soaking (2), roasting pan (3), and sun drying (4). [*Courtesy*: James Wimberly, Ford Foundation].

An yet another instance of the traditional and widespread use of this principle is an intermediate product transiently appearing during the production of flaked rice and puffed rice in the Indian subcontinent. In the commonly practised methods, paddy is first soaked and then roasted to produce an intermediate product which, unknown to the researchers or producers, is nothing but dry-heat parboiled paddy (see detailed description in Chapter 8, Rice Products). The processed paddy is then either pounded or pressed (flaked rice) or milled and puffed (puffed rice). It is only when CFTRI researchers studied these production processes, the fact of production of dry-heat parboiled paddy/rice as an intermediate product here came to light. These processes have apparently existed for centuries.

The dry-heat process was thus already prevalent in isolated pockets before its potential was understood in laboratories. It became known only when scientists were trying to study certain products (flaked rice) or trying to develop rapid systems of drying of paddy by direct heating. Now the process is well understood and the special properties of the product are well known.

In fact, once understood, a production process was developed in the CFTRI using a 'gram roaster' (Fig. 7.14). This is a rotating conical drum containing two spiral channels inside, one for sand going backwards and the other for a mixture of sand and paddy (or any other material to be roasted) going forward, the whole heated from below (Fig. 7.14). Soaked paddy is fed at back, gets mixed with hot sand and moves forward, comes out roasted into a screen, from where the retrieved sand re-enters the roaster and moves backwards. This is a simple, efficient system for dry-heat parboiling. But it does not seem to have evoked much of an interest as a means to produce parboiled rice. Interestingly, in contrast, the same process has generated considerable interest among producers of flaked rice and is being widely adopted for this product wherever the scale of production demands.

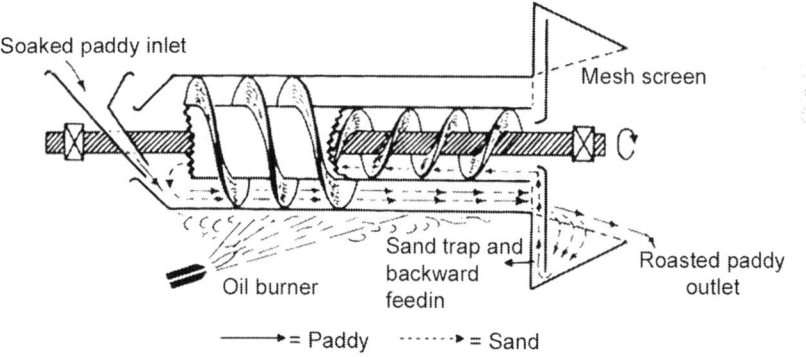

Figure 7.14 Schematic diagram of a 'gram roaster' [Reproduced with permission from Ali and Bhattacharya (1980)]

Although sound in principle, the concept of dry-heat parboiling as a means of producing commercial parboiled rice seems to have limitations. Other than the 'stove process' in Brazil, which was discovered by accident and got established, this method is not known to be used anywhere else now. So the process seems to have some limitations for large-scale production.

7.4.6   Brown-rice parboiling

If the husk of the paddy generally appears as an unnecessary encumbrance while trying to prepare a rice product, it is often forgotten that the husk provides protection against damage in any general handling and processing of rice. It is nonetheless true that if any process can be developed which directly works on the brown rice but can satisfactorily avoid damage, then it is bound to be simpler and cheaper. This is especially true in parboiling. Brown rice is approximately two-third of the volume and four-fifth of the weight of paddy. In addition, brown rice hydrates probably ten times faster than paddy. The advantage of processing of brown rice rather than paddy for parboiling is thus obvious. However, being clearly much more delicate, handling of brown rice is more difficult and risky than that of paddy.

A process of direct parboiling of brown rice has been developed. Soaking of brown rice is simple and can be achieved in 1–2 hours at 50–60°C. It can be done in a reasonable time (<10 hours) even with ambient water. But cooking and then its handling have to be done with care so as not to damage the grain. One possibility is conduction heating with sand. This has been achieved using a gram roaster in pilot scale in the laboratory. But the process is not known to be in use. Alternatively, superheated steam or high-temperature air can be used. A company in USA is said to be using the process in one plant but the details are not known.

7.5   Present status of the parboiling industry in the world

Despite its overwhelming concentration in south Asia, parboiling of rice is quite an extensive industry in the world.

An approximate estimate of the status of the production and international trade in parboiled rice during the turn of the millennium is shown in Table 7.1. Small amounts of parboiled rice are fairly widely produced in several countries outside south Asia. But only a part is often for internal consumption, if at all, and the remaining for international trade. Parboiled rice has many interesting properties. It can make good puffed rice, canned rice, quick-cooking rice and cooked rice in pouch. For making good *pilaf*, parboiled rice is better suited. For cooking rice by inexperienced people, parboiled rice is a safer bet. Besides, head rice yield is much more after parboiling. And nutritionally it is superior. For all these good qualities, parboiled rice is an important commodity for international trade. A substantial part of it in many countries is therefore produced only for trade. This is mostly true, for example, of Thailand, USA and a part of the production in Europe (Italy, Spain, France). Apart from that, parboiled rice is usually cheaper than raw white rice (yield is more). So, many low-income food-shortage countries prefer to import ordinary-quality parboiled rice. The only place other than south Asia where a fairly substantial amount of parboiled rice is produced only for internal consumption is Brazil in South America. Some parboiled rice is also produced by primitive systems for internal consumption in many African and Caribbean countries. Interestingly, these practices usually originated with the initial arrival of Indian immigrants into these countries.

Table 7.1 Approximate annual production and trade (million tonnes, milled equivalent) in parboiled rice in the world (around 2000 CE) [Reproduced with permission from Bhattacharya (2004)]

Area/Country	Production	Export	Import
Asia			
India	50	1	–
Bangladesh	23	–	0.5
Sri Lanka	2	–	0.2
Thailand	1.8	1.8	–
West Asia	–	–	1
North America			
United States	0.8	0.5	+a
Central and South America and Caribbean	2.4	+	+
Brazil	2	–	–
Africa			
West Africa	+	–	1.5
Southern Africa	–	–	
Europe	0.6	+	0.2
Italy	0.3	+	-
Spain	0.15	+	–
Total	81	3.5	3.5

aThe + sign indicates a definite but minor quantity (<100,000 t).

Other than south Asia, Africa and Caribbean, production in the rest of the world is mostly made by one or the other modern or semi-modern processes. Production systems in Europe are obviously by modern techniques carried out in fairly large installations ranging in capacity from 2 to 20 tonnes of paddy per hour. These are located in Italy, Spain and a few in France. The systems are either Gariboldi technology or its variant or increasingly more now the so-called American system. The product is mostly used in industries producing different rice products (quick-cooking rice, canned rice, rice in pouch…) or for export to each other or to central and east European countries. Production in USA is entirely by highly sophisticated technologies and the entire production is carried out by only a handful of companies. The output is used by various industries or for export. The parboiling industry in Thailand is entirely for export. Semi-modern, modern and highly sophisticated technologies are in action depending on the desired product and its destination. The production in Brazil is entirely for internal consumption. Part of the production is by the stove process mentioned above and a part by semi-modern processes (variants of CFTRI technology). Small amount of parboiling is extensively carried out in African and Caribbean countries for internal consumption. These are mostly by primitive or slightly improved technologies.

Saudi Arabia, Iran and the nearby countries of West Asia are big importers of high-quality parboiled rice, including parboiled basmati rice. High-quality products are also traded among European and North American countries. Substantial quantity of consumer-quality product is imported by the Caribbean countries and Bangladesh. There has been a recent spurt in import of parboiled basmati rice in West Asia, including Iran. With this increase, probably 6–7% of world's rice production is internationally traded.

As mentioned earlier, the overwhelming majority of parboiled rice is produced and consumed in south Asia, including Bangladesh and Sri Lanka. The industry in India is now in an intermediate stage of modernisation. Primitive technologies are still prevalent, but the modernised double-boiling process is steadily being adopted. The industry elsewhere in south Asia is still largely in a premodern stage. But hot-soaking or 'modernised double-boiling' processes are making steady progress.

7.6   Properties of parboiled rice and effect of processing conditions and rice varieties

Parboiled rice is no doubt rice, but its properties are quite different in many respects from raw white rice. Overall, we can say that the following are the ways in which parboiled rice differs from raw rice. This list will also give an idea about the properties of parboiled rice.

- Milled raw rice grains can be considered as relatively opaque and white. In contrast, parboiled rice grains are relatively speaking glassy, translucent and with a light yellow-brown colour.
- The oil content of the rice grain, even though quite small, is largely located in the outer layers of the grain, especially the bran layers. That is why rice bran is a reasonably good source of oil (15–20%). Parboiling process has the character of pushing the grain oil even more outwards. As a result, the bran from parboiled rice contains more oil (20–25%). The milled rice also tends to have a little more oil on the grain surface. As a result, the rice tends to have somewhat poorer flow and packing properties (more tacky).
- The pre-existing defects of cracks and chalkiness in rice are completely

healed by parboiling. This is the reason why the breakage of rice during milling is dramatically reduced after parboiling. Parboiled rice therefore always gives much more head-rice yield than raw rice. This is the reason why parboiled rice is generally cheaper than raw white rice.

- The vitamins and certain other micronutrients are mostly located in the outer layer of the rice grain. Actually this is true of all cereal grains. As a result of this location, most micronutrients including Vitamin B are lost when the rice grain is milled, specially overmilled. Parboiling process pushes or fixes some of the most important nutrients inside the rice grain. Therefore, vitamins are not extensively lost during milling of parboiled rice. One may recall that this is the property which brought parboiled rice into the notice of the world as a preventive for beriberi.

- The cooking–eating property of parboiled rice is quite different from that of raw rice. Cooked parboiled rice is much more firm and discrete than cooked raw rice.

Two more points are to be noted in this connection. The first is the effect of parboiling conditions on the product quality. One can notice from the earlier discussions that parboiling can be done in various ways. The product quality differs to some extent accordingly. For example, pressure-parboiled rice is much more hard and discoloured than normal parboiled rice. It is even difficult to cook. On the other hand, dry-heat parboiled rice has the ability to absorb water even at low temperatures. Another source of variation is the degree of parboiling. One can easily see that the severity of the process of parboiling can vary between fairly wide limits. Thus soaking can be done between room temperature and 70–75°C. Steaming can be done for different lengths of time under different pressures. We can easily say that these variations in processing conditions are bound to affect the product quality. The more severe the heat treatment, the more hard and discoloured the product. By adding a pre-steaming step before the soaking and steaming process, we can also add to the state of age of the resultant product. We can thus produce a product with a wide range of qualities by varying the conditions of the process.

A second source of variation in product quality is the raw material, rice. There is inherent variation in rice quality depending primarily on its amylose content. As already mentioned, low-amylose rice (prevalent in east Asia) cooks soft and sticky. High-amylose rice is prevalent in south Asia and cooks firm and fluffy. Intermediate-amylose rice (prevalent in southeast Asia) cooks intermediate. It is a matter of great interest that these differences in cooking–eating quality are carried over even after parboiling. Thus, other things (i.e. processing conditions) being equal, a low-amylose rice would yield softer and stickier cooked rice even after parboiling than parboiled high-amylose rice. Although parboiling, generally associated with south Asia, is by and large done only with high-amylose rice, the above property is theoretically a source of variability in parboiled product and can be adopted by industries if and when desired. By combining the rice type and processing conditions, therefore, and again so with the age of the rice as a variable, one can clearly produce an endless variation in product quality in parboiled rice.

Further reading

Various aspects of parboiling of rice, including its methods of production, properties of parboiled rice and its trade, have been extensively discussed in the following:

- Ghose et al. (1960) – Description of the then situation, particularly in India
- Gariboldi (1984) – Summary of situation as then known
- Bhattacharya (1985) – Science and technology of parboiling
- Bhattacharya and Ali (1985) – Chemical changes occurring in grain during parboiling and properties of parboil rice
- Pillaiyar (1988) – Science and technology of parboiling
- Luh (1991b)
- Bhattacharya (2004) – Science and technology of parboiling
- Bhattacharya (2011) – Theory of parboiling and properties of parboiled rice
- Bhattacharya (2013b) – Continuous steaming system for soaked paddy

Rice products

8.1 Introduction

Cereals are the staple foods of humankind. Cereals used to provide in olden times the bulk of our food intake, accounting for most of the energy and a substantial part of all other nutritional intakes. Now their relative contribution to our nutritional requirement is gradually decreasing. Yet cereals are still the major source of food, energy, even protein and other nutrients for humans, other foods providing variety and supplementation for the overall nutritional need. This is equally or more true of rice which, among cereals, is the most important staple food of humankind, accounting for the food requirement of more number of people than by any other food grain.

8.1.1 Need for rice products

The bulk of the intake of rice as food is as table rice. Rice is cooked by boiling in water and then served. In the 'rice country' (comprising of south, southeast and east Asia), where over 90% of the world's rice is grown and consumed, people have been eating rice in this manner at least twice a day for centuries as their main meals. However, apart from the two main meals, human beings also need to eat breakfast and other snacks as subsidiary meals. Here again cereals, including rice have had to largely fill this gap. But rice for this purpose is not generally consumed in the form of cooked table rice but in some other modified form. These are the forms that may be called rice products. The objective is to break the monotony of the food by adding variety and surprise. Rice therefore has had to be adapted to provide for the needs of breakfast and snack foods as well.

8.1.2 Regional diversity in type of rice products

Human ingenuity applied over centuries has helped produce numerous products from rice. But there is a curious regional difference in the kind of the products that are made. Among the rice countries of Asia, the majority of the products made in southeast and east Asia are cakes, pancakes, fries, puddings and allied foods made from wet-ground rice with or without other additives and then cooked. In south Asia, on the other hand, the situation is a little different. It is not that rice cakes and fries are not made here, but these do not dominate the scene. The three main products made here are flaked, puffed and popped rice prepared not from dry- or wet-ground rice, but from whole-grain rice. This is quite opposite to the situation prevailing in the other two regions. It is again not that puffed and flaked rice are not made in southeast and east Asia, but it is cooked cakes and fries made from wet-ground rice that dominate the scene. As if as an exception which proverbially proves the rule, a fermented and cooked rice cake, *idli,* is widely produced in some parts of the south- Asian region. However, be it noted that *idli* was originally confined only to the southern part of India, although now it has spread more or less throughout that country.

We may extend this reflection on regional difference in product type to the non-traditional West. One may note that breakfast and snack foods from rice have of late been introduced into the wheat-eating areas of the world as well. This happened due to a variety of reasons. One reason has of course been the fact that the world has today become much more integrated than before. People are today willing and eager to experience and taste and exchange foods and other cultural artefacts from different regions. Products from rice have thus added variety to peoples' diet in the West. But the more important reason has been on health considerations. Wheat is well known to cause an allergic response, celiac decease, in a small number of people. It is generally considered that the protein of the wheat, viz. gluten, is responsible for this allergic response. Be that as it may, people suffering from celiac disease have had to be provided food made from cereals other than wheat. Rice came very handy in this context. Rice is well known to be largely non-allergenic; therefore foods and snack foods made from rice were well suited to fill the gap in such cases.

There may well be one more curious reason for the entry of rice-based breakfast and snack foods in the Western diet. There has been a gradual uptrend in the international trade in rice for some time. Now, on an average, approximately 15% of broken rice is produced during modern milling of rice. In the traditional rice mill output within the traditional rice countries, this proportion of broken rice would not normally be considered as a big issue and would not in fact necessarily be separated from the output. But such would not be the case in the West. As per the strict quality standard norms prevalent in modern industrial societies, the broken rice would have had to be separated. An outlet other than as table rice had therefore perforce to be created for this broken rice. Production of various rice products, including pet food, naturally arose as a consequence. It should also be noted that this production then automatically evolved into a modern industry.

Again, in this industrial environment, much of the new products had necessarily to have been made from rice flour. Rice flour itself too thus became a kind of another rice product and it evolved into an item of industrial production and commerce, for being converted into various products in downstream industries. This scene thus provides another case of regional difference in the kind of rice products – modern industrial products, including rice flour, in the West.

To summarise, we can say, we have largely (a) traditional cooked cakes and fries made from wet-ground rice (largely waxy rice) in southeast and east Asia, (b) whole-grain traditional products made from usual non-waxy rice in south Asia, and (c) modern industrial products made largely from dry-ground flour in the West.

8.1.3   The products classified

Reliable figures showing how much of the rice crop is made into various products in different countries are not generally available. In India, it is commonly believed that 10% of the rice production is made into various products. How reliable is this estimate is not known. Whatever it is, a similar proportion should be made into products in the other rice countries of the Asia. The proportion might even be more if one includes the production of noodles. Some rough estimate of the production figures is available from USA. It has been estimated that 40–45% of the rice output in USA is exported, leaving 55–60% for domestic use. About two-thirds of this is used as table rice, leaving about one-third for production of rice products. But one should note that more than half of this one-third is used for making pet foods and for brewing, leaving a relatively small proportion for breakfast and snack foods.

A list of the more important rice products and a scheme for their classification are presented in Fig. 8.1. The products have been mainly divided into two groups – traditional products in Asia and modern products in the West. Obviously, the latter products form only a minuscule proportion of the total quantity of rice food products produced and consumed in the world scale. But obviously, historically speaking, slowly but steadily these are gaining ground and may even dominate the scene in some not too distant future. A third group constitutes, what may be called, for lack of a better term, unusual industrial products from rice.

In what follows is presented a brief description of the more important ones of the numerous rice products. The modern industrial products prevalent in the West are taken up first, followed by the numerous traditional products prevalent in the rice countries of Asia. The unusual industrial products are discussed last.

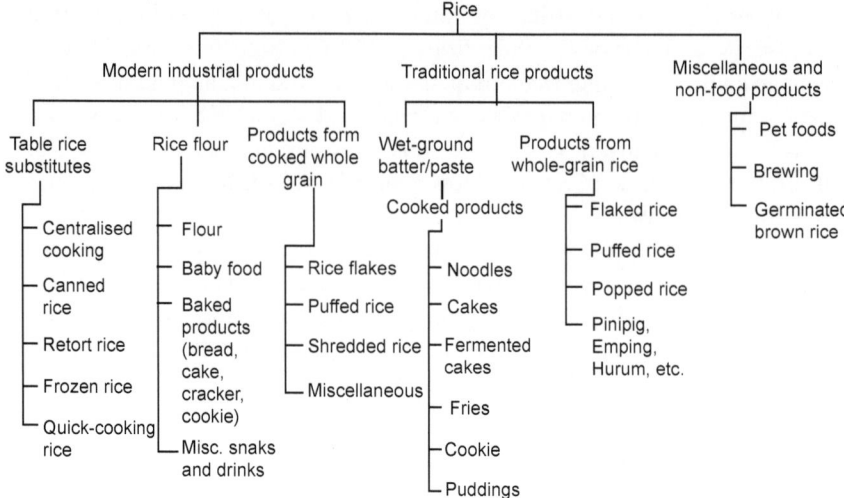

Figure 8.1 Rice products and their classification

8.2   Modern industrial rice products: Cooked table rice substitutes

Cooked, whole-grain rice is the form in which rice is overwhelmingly consumed. Cooking for this purpose is almost universally done at home. But with change in modern life circumstances, various substitutes have appeared in the scene whereby the need for cooking of the rice at home is being avoided or at least minimised.

8.2.1   Distributed cooked rice

In rice-eating societies where modern technological production is in wide use, centralised cooking and distribution of cooked rice become a feasible proposition. This is particularly true when a large number of people work in big organisations employing large groups of people, or numerous students gather in schools and colleges. Among rice-eating societies, this description fits Japan well. Indeed such a system of centralised cooking of rice and its transport and distribution to various target groups of people working or studying together has been in operation in Japan. Basically for such a system, the primary arrangement has to be a centralised cooking system combined with efficient, fast and hygienic distribution system. The system comprises of a centralised production facility where rice along with other necessary ingredients are being efficiently and optimally cooked in a continuous process. The material is then packed in appropriate containers and then moved through the public transport system to various destinations and distributed there.

There is nothing necessarily very great with respect to a high-tech technology being involved here. The hallmark of the system would be the foolproof integration and management of the entire process which is a sum total of a large number of steps of production, transport and distribution. Obviously there is no special requirement of any raw material rice in this instance. Whatever the variety and type of rice that the target consumers are accustomed to or have the liking for, would be used in the process.

The Japanese example has been cited here. However, although feasible and apparently appropriate in theory, a widespread use of such technology in the world, applicable if at all only in rice-eating societies, seems to lie in the future.

8.2.2   Canned rice

Let us now move away from centralised production and distribution of ready-to-eat rice and food for a large number of people eating together in institutions under a single roof. Ways of replacing of cooking of rice at home seem more feasible. Thus cooked rice can be produced and distributed for individuals or families eating at home in different ways. Small quantities of rice can be cooked and put in cans or pouches or cups.

Canned rice was one of the earliest rice products put in the market in industrial societies. These attempts have a history of nearly a century. The first patent on canned rice was taken out in the 1920s. The material has a chequered history. Initial attempts were obviously made with raw white rice and the results were invariably not very optimistic. When news of parboiled rice reached Western rice industries around World War II, the process took a step forward. Parboiled rice appeared much more suitable for making canned rice as compared to raw rice. Several patents on parboiling were granted and a modern parboiling industry sprung up in the West. This came as a boon to the canned rice as well as some other rice industries. Cooked parboiled rice was found to remain firm and whole and much less frayed during cooking. This cooked rice therefore stood up to the punishment of retorting much better than raw rice. When this was combined with the action of various salts added during cooking (sodium chloride, sodium sulphate, calcium chloride, magnesium salts were tested), as a means to reduce the loss of rice solids into the liquid, further improvements in the product quality were seen. Other step-by-step improvements in the production technology followed, and reasonably satisfactory products were being marketed. However, other events caught up with the industry, especially the availability of newer products in the form of rice in pouches.

The main problem in preparing canned rice is two-fold. One is mashing of the rice, including splitting of the grain and fraying of its edges. Unless the rice is strong enough to stand up to the stress of retorting, such mashing and deshaping are inevitable. The second problem is loss of solids during cooking and especially their entry into the canning liquid. Any loss of solids is a direct loss to the industry. But, additionally, loss of solids into the canning liquid creates unfavourable consumer response. Any movement of particulate solids into the canning liquid or soup make the liquid turbid and hence unacceptable.

Efforts made to overcome these deficiencies were four-fold. First were fine adjustments of manufacturing conditions to improve the efficiency of the process. The second as mentioned was the introduction of parboiled rice. This step immediately upgraded the product significantly because of the well-known superior cooking quality of parboiled rice. Still, this was only an improvement and not yet entirely satisfactory. Then came a well-coordinated research both in USA and France to study the effect of rice varieties on canned-rice quality. The conduct of a well-coordinated study on the effect of various characteristics of rice (including its amylose content, gelatinisation temperature, protein content, water absorption capacity, fat acidity, starch-iodine blue value, paste viscosity, etc.) on the cooking-eating quality of rice added further impetus to the efforts during 1950s and 60s. It became clear for the first time that various factors in rice, the most important among which was its amylose content, played a big role in determining its cooking-eating property including its texture. The relation of these various factors to canned-rice quality was then meticulously ascertained. Use of appropriate varieties based on these investigations further improved the quality of the product.

The fourth stage was an attempt to strengthen the rice by chemical treatment. Cross-linking of starch in the rice with chemical agents such as phosphorous oxychloride or epichlorhydrine improved the structural integrity of the rice, leading to a great improvement in product quality. However, legal approval of the process has been pending and hence this approach could not be utilised.

Meanwhile, another recent innovation in parboiling technology has a potential in this respect. It has been recently shown that 'pressure-parboiled' rice produced under certain conditions has an excellent ability to stand up to the stress of retorting much better than conventionally produced parboiled rice. This process should be of benefit to the canning and retort industry.

8.2.3   Retort rice (rice in pouch)

The other form of distribution of cooked rice is in the form of rice in retort pouches. After the development of flexible heat-sealable pouches, the system of canning of cooked rice or of other food is probably in the way out. Canning of food is gradually being replaced by production in pouches. The principle is largely the same as in canning. Either the rice is cooked and put in pouches or rice and water are taken in pouches and cooked. The pouch is sealed and suitably retorted. Use of retort rice has been rapidly increasing in recent times. The convenience of the product for single persons or for small families for dinner is obvious. All that is needed is to warm the pouch in hot water for a few minutes or in a microwave oven (after puncturing the pouch) and the food is ready for serving. The versatility and convenience of the product are obvious. Not only plain rice but also the production of a complete meal by mixing with proper sauces, spices, vegetables or meat is feasible.

Here again the rice (and other ingredients if present) is susceptible to structural damage during retorting. Appropriate process variables, including the use of properly parboiled rice, is therefore necessary.

8.2.4   Frozen rice

Use of frozen rice is another way to distribute cooked rice ready for consumption as table rice or in mixed cooked form. In this system, rice is cooked and frozen. It is then preserved and distributed in frozen condition. Frozen rice, or the mixed food, is heated in the microwave oven in the target centres (homes or chain restaurants or dining car or other eating centres) and served. Rice alone or rice with various combinations of sauces, vegetables or meat can be made this way.

8.2.5   Quick-cooking rice

This product does not completely fall in the category of ready-to-eat rice or rice food cooked outside the home. Here the need of cooking at home is not entirely eliminated but its burden is reduced. Rice normally needs some 15–25 minutes of boiling in water to cook (if it is raw white rice) and at least one and half times of this duration if it is parboiled. In quick-cooking rice, the rice is precooked and then dried in a production unit in such a way that it can cook in 5–10 minutes at home. This concept too arose from around the middle of the last century. The first patent was granted to the General Foods Corporation in the USA in 1948. Since then, newer patents have been granted off and on and all along further improvements have been regularly made. The basic principle of these processes lies in precooking the rice in water, followed by drying it in such a way as to leave a large number of cracks or openings in it. The semi-porous structure allows channels for entry of water to enable the product to be cooked quickly. Alternatively the grain is micro-damaged by suitable dry heat or freezing.

One system is the soak-boil-steam-dry method. In this method, milled rice is soaked, partially cooked, then steamed, again cooked and so on until it attains a moisture content of about 70%. The cooked rice is then carefully dried so as to maintain a fissured structure. Alternatively, the partially cooked rice is bumped or rolled (gently pressed between two rollers), i.e. lightly flattened, and dried. The increased surface area enables the product to cook faster.

In another approach, advantage is taken of the principle that cooked starchy material can be puffed by dry heat and such puffed matter hydrates or cooks quickly. So in this approach, rice is cooked and bumped as above. The bumped grains are then lightly toasted to generate partial puffing which helps in quick hydration. Alternatively, the grain is gelatinised, cooked and then dried. The dense, glassy cooked grains are lightly puffed to make a porous structure which hydrates quickly. Yet again, milled rice itself can be adjusted to a proper moisture and then lightly gun-puffed. The resulting porous material hydrates quickly and becomes quick-cooking rice.

In yet another alternative approach, the structure of milled rice can be lightly damaged by a suitable treatment to enable it to absorb water quickly. Thus, milled white rice or brown rice may be heated in air to develop fissures in the structure. These fissures can enable the rice to hydrate rapidly by allowing entry of water and thus to cook quickly. Alternatively, raw or cooked rice can be frozen and then freeze-dried. The resulting fissured structure enables the grains to cook quickly.

Recently an entirely new approach has been introduced. In this process (details of which are not easily available), parboiled rice is directly used without any further cooking. It is said that paddy is parboiled and then milled by a special process when still somewhat moist. This milling under moist condition is said to produce internal stresses in the grain resulting in micro-cracks within the internal structure that are not visible from outside. These cracks are said to enable the rice to hydrate and cook quickly. This is a major innovation, for the rice looks no different from usual milled parboiled rice. In all the other processes described above, the dried product looks perceptively different from normal milled raw or parboiled rice. While this altered appearance may or may not be considered as a deficiency, depending on one's psychological perception, the fact of it having gone through a processing treatment is obvious to the eye. But that is not the case in the latest process being discussed here. In this instance, the product is indistinguishable in appearance from normal milled parboiled rice. This is a major plus point for the product. Even though this perceived 'benefit' is only a matter of psychological perception, the marketing value of this perception is clear.

8.3   Modern products: Rice flour and its products

8.3.1   Rice flour

Rice flour is now-a-days considered a fairly important product from rice. Strictly speaking, rice flour is not a product in the same sense as, for instance, rice cake or rice bread or rice cookie is. It is not something that a consumer can purchase and heat or boil or cook and eat. It is at best an intermediate product that has to be further worked upon to make something that can be consumed. However, rice flour is an important commercial commodity today which is manufactured and fairly widely traded and used for production of various rice products in industrial societies. This is especially more so after the recent WTO international trade agreement on agricultural commodities.

Many products are now being made in industrial scale using rice flour as a starting material. This is especially the case in the West and partly in Japan where breakfast, snack and baby food products have lately become common industrial products. Rice flour is now being widely used in West for production of baby foods, breakfast cereals, snack food, extruder products, and baked foods such as bread, biscuit, cookie, pizza and so on. It is also widely used as dusting flour for separating dough pieces, for pan release and to impart crispiness. As mentioned earlier, these uses have especially arisen because of the need to cater to the requirements of a small section of wheat-eating people who are allergic to the wheat gluten. People suffering from celiac deceases must be provided for by alternative sources of their food requirement other than wheat and its products. Rice and rice products are ideally suited for such a situation for rice is well known to be a non-allergenic food. It is also for this same reason that rice is often used as the first cereal to be used in infant food.

Rice flour is regularly made at homes and catering establishments in traditional rice-eating Asian countries too. But that is not as a separate product at all. It is only as an intermediate material, mostly in the form of a batter or a paste. Here it is only produced as a part of an integrated process of making various traditional home products such as rice cakes and noodles.

Grinding, usually wet grinding, of rice is most often the first step in the preparation of such home-made rice snack products. In that sense these may also be considered as some sort of production of rice flour, but that is not as a separate product but as a part of an integrated process of preparing the rice food.

Rice flour as an independent product is usually made by dry grinding. It is largely manufactured from broken rice. As an intermediate product, in the form of a batter or paste, for making a traditional rice food, no distinction is made between whole and broken rice and both or their mixture are used without any distinction. That is precisely because the intention is not to produce flour but a product and the batter or the paste is only an intermediate product. In the case of producing rice flour as a commodity for trade, on the other hand, broken rice is what is most commonly used and quite understandably so. A sizable amount of rice is normally broken during the process of rice milling (about 15% or more) and preparation of rice flour is one of the recognised outlets for utilising this broken rice. Second head rice, i.e. pieces of rice approximately half the length of a full grain, is most widely used for this purpose. Brewers' rice (less than half grain size) is not commonly used for this purpose. This fraction generally contains more stones and other impurities and is difficult to clean thoroughly. It is therefore usually avoided for making an edible product.

Another advantageous point in favour of rice flour is its versatility. (i) First, it is well known that rice varieties differ widely in their intrinsic character, based mainly on their amylose content. Therefore one can produce rice flours of a range of properties using rice of diverse qualities. (ii) Second, the flour in theory can be prepared either by dry or by wet grinding, leading to products having somewhat different characteristics. (iii) Then there are different types of grinders available. Use of these different grinders produces flour of somewhat different properties, including production of different particle sizes and having different extents of starch damage. (iv) One can also use different forms of rice for grinding. Use of either raw rice or parboiled rice obviously produces lots having different properties. (v) Again one can use either unpolished brown rice or milled rice. Flour from brown rice would impart a different flavour and texture to the product. (vi) Age of the rice being used for grinding is another factor. The properties of the product would surely show some difference depending on whether the rice being used is fresh or aged. Clearly, rice flour has a wide versatility in terms of use for different products rarely encountered with any other cereal grain. One may also add that starch granules of rice are among the smallest in size (3-10 μm), which imparts special properties including use as a fat replacement.

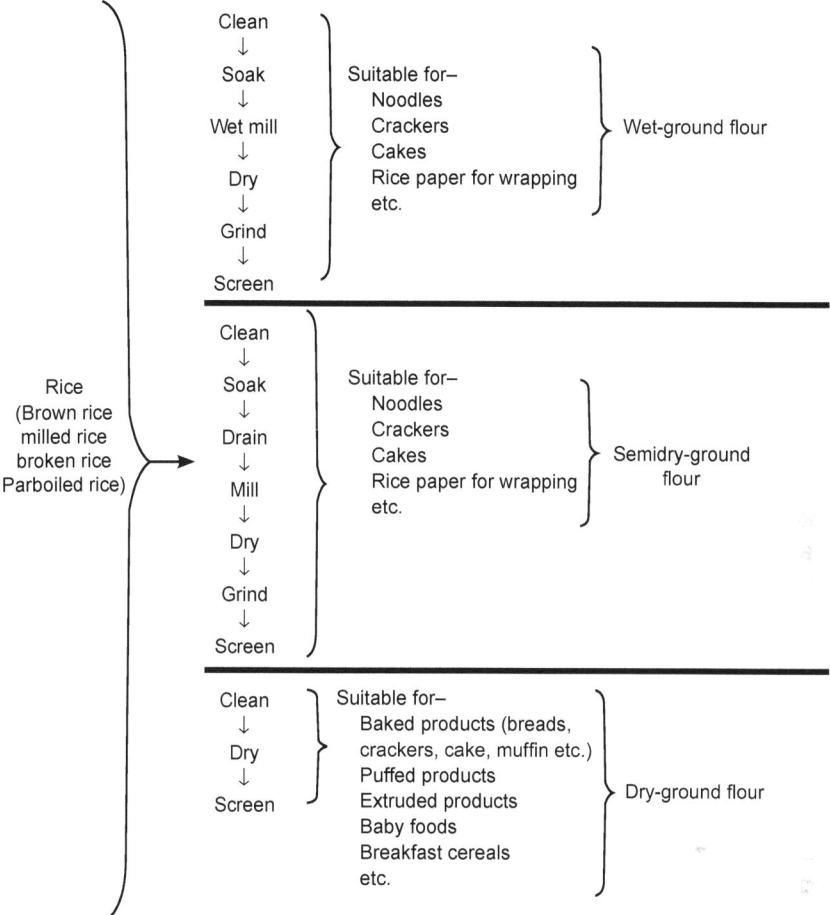

Figure 8.2 Manufacture of rice flour and their applications

Even if some of the factors mentioned above produce only a rather minor difference in flour properties, some special differences would be obvious. One is the varietical difference among rice related to amylose content. That flours from rice varieties having different amylose contents, such as high, intermediate, low and negligible amylose, would behave differently and would produce products with rather widely different characteristics is quite obvious. For example, many products would appear harsh, dry and sandy if made from high-amylose rice. On the other hand, some other products, e.g. noodles, cannot be made from low-amylose rice. This difference is more explicit in waxy rice flour. Waxy rice flour, with no or negligible amylose, has excellent thickening property for sauces, gravies, puddings, etc. It is well known that waxy starch is resistant to retrogradation and waxy rice starch is least susceptible to retrogradation among all waxy starches. Its paste can therefore undergo several freeze-thaw cycles without syneresis (i.e. liquid separation). This is a unique property and is the reason why waxy rice is used in many manufactured products, such as sauces and gravies.

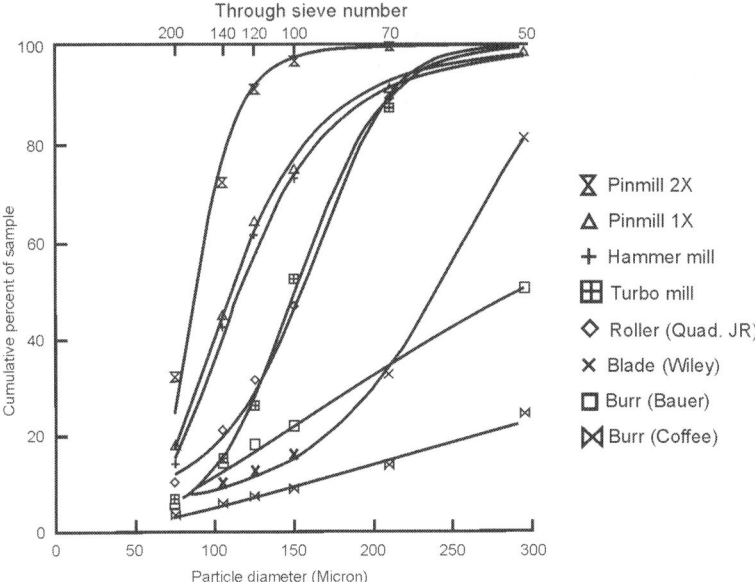

Figure 8.3 Cumulative particle size distribution by weight for rice flours ground by various mills [Reprinted from Nishita and Bean (1982)]

Rice flour can be prepared by wet, semi-wet and dry grinding methods (Fig. 8.2). As already explained, wet grinding is mostly used for making traditional products, usually in the form of a batter or paste as part of an integrated process for making the products. Flour *per se* is usually not made by wet or semi-wet process. Wet grinding has the advantage of reduced starch damage. The presence of water helps in cooling as well as providing lubrication during grinding. This cooling and lubrication help in preventing large-scale damage of starch granules. Unfortunately the extra cost involved in drying and in disposal of the waste water prevents the wide use of this technology except in some niche areas. The cost is minimised in semi-wet grinding. But the process is such that it is probably more suitable for use in small- or cottage- scale enterprises. Dry-ground flour no doubt has more starch damage but this is unavoidable. It can be used for baked products and baby food and extruded products. The effect of starch damage can also be overcome or minimised by use of certain techniques, such as high-speed mixing.

In dry milling, different grinding equipments produce flours having different particle sizes (see Fig. 8.3). Turbo and hammer mills produce very fine particles but also cause much heating resulting in more starch damage. A roller grinder is best for baking. A flour having a particle size such that 50% passes through a 100-mesh screen is best for baking. Finer flour has more starch damage. Certain products require precooked or pregelatinised starch. Flour made from parboiled rice or quick-cooking rice may be suitable for such purpose. Again brown-rice flour imparts certain special properties to the product. The texture of the resultant product is more chewy, which may be liked in certain products. However, it is also true that brown-rice flour may cause development of free fatty acids due to the presence of the enzyme lipase unless the material is properly heat treated. One way out of this difficulty is to add stabilised bran to milled rice flour as a substitute for brown-rice flour.

High-protein rice flour is another potential product and can be of value in making infant food. It can be prepared in different ways. One way is to carry out gentle and gradual abrasion of the outer layers of the milled grain which contain very high levels of protein. It is possible to obtain rice flour with up to double the normal level of protein in this manner. Another way is to use air classification of normal rice flour. Alternatively, high-protein rice flour can also be made by enzymic digestion of a part of the starch in regular rice flour.

8.3.2   Baby food and infant rice cereal

Traditionally various staple cereals in combination with legumes and other foods have been suitably cooked and mashed for use as the first solid food for infants. This has been a well-worn custom and is still being practised in most parts of the world. Inevitably in the West, to start with, and now in many parts of the world, especially in modern urban areas, a baby food industry has come up to provide for a ready-made first solid food for the infant. Its advantage first and foremost is convenience. Secondly, it provides a nutritious food of standard characteristics. It also provides a simple method for including various micronutrients in the infant's diet. Thus it is a simple way to provide for adequate intake of iron and other minerals and vitamins in the infant's food. As rice is well known to be hypoallergenic in character, it is most commonly used in this precooked infant cereal.

Certain characteristics are normally expected in precooked infant cereals. The first requirement is easy reconstitution of the food. The cereal must mix easily with milk without forming lumps. Also important is the amount of liquid required to prepare the food. If the amount of liquid required is too high, the parent would feel that the baby was not getting enough solids. On the other hand, if the required liquid is too little, then the volume would look too small. The same is true of the viscosity. If the reconstituted food is too thin, then the food would look insufficient. But if it is too thick, then it might be difficult for the baby to swallow. Diastatic enzymes are often added to the mixture to reduce viscosity. The purpose is not only to make the food easy to handle but also to increase its calorie density.

Manufacturing of infant food cereals is a highly skilled job which is as much an art as a technology. The basic principle is to make a slurry with the powdered ingredients including the cereal (mainly rice) flour. The slurry is optionally lightly cooked. It is then dried in an atmospheric drum dryer. The drum drying step requires a lot of experience. The operation must of course cook the material properly. But the output product must not be too thick or too thin. It must also have a correct bulk density and be able to be easily scraped off the drum. The dried sheet is then lightly ground to produce the flakes. The composition and concentration of the slurry, its precooking, the speed and temperature of the drum, the consistency of the paste and clearance between the drums, must all be right to obtain the best product. It needs hardly to be mentioned that the process is very versatile providing an opportunity for using diverse materials including micronutrients as needed.

The raw material rice must also be properly chosen. As already mentioned, second head broken rice should ideally be used. Smaller brokens are avoided because these are more likely to contain impurities and other seeds. Among the rice types, low-amylose and low-GT (gelatinisation temperature) rice are more commonly used (short- and medium-grain rice in USA). The lower amylose content helps in reducing retrogradation. The low GT on the other hand helps in reducing the energy use in manufacturing. In addition, the lower GT is beneficial while using diastase enzymes, for then the enzyme is more likely to remain active even when the rice is pasted.

Drum drying has been the traditional method of preparing infant foods. But these days extrusion cooking is being more and more used because of its versatility. However, the basic principles remain the same. Not only different cereal staples but also a diversity of other ingredients can be used with ease.

8.3.3   Baked products

Baking has been one of the common methods of preparing food from cereals throughout history. This is especially true of wheat and its products. Wheat is unique in having a special protein, the gluten, which forms a network when mixed with water. This network structure helps to retain the fermentation and other gases and prevent their escape during heating. It thus enables the structure to rise (leaven), thus forming a light and porous product, so well illustrated by the loaf of bread. Most other grains, including rice, do not possess a gluten and therefore do not easily leaven and hence are not suitable for making bread. However, baked foods from rice are required to be made primarily with the object of providing appropriate foods for those suffering from gluten allergy. Other circumstances too have promoted use of rice in baking, such as the need to add variety to people's food, and the need to provide for an outlet for the broken rice. Further, there are countries where baked foods have been introduced but enough wheat is not grown to meet the requirement. Use of other grains, especially rice, as partial replacement of wheat, becomes unavoidable in such circumstances with the idea to extend the availability of the wheat grain. It needs hardly to be mentioned that baked products from rice (or other cereals) inevitably largely mimic the existing baking-industry processes and products from wheat. The baking industry using wheat is so well established that baking of rice or other cereal products inevitably follow the well-trodden path of the existing wheat-based industry with modifications as are unavoidable.

Rice lacks the gluten protein of wheat and what protein the rice has does not develop an interconnected structure upon mixing like gluten and therefore cannot hold fermentation gases. As a result, rice cannot be directly substituted for wheat in an yeast-leavened product. At the same time, some method of holding the gases to enable the product to rise during heating is essential for baking. Consequently, the ability of various gums and surfactants to act in this manner in holding gases has been extensively studied. Only one gum, hydroxypropyl methylcellulose, has been found able to retain gases in this manner. This gum has been found sufficiently effective for preparation of an acceptable bread.

Rice grain types also affect the quality of the bread. Only US short- and medium-grain types have been found suitable to give an appropriate soft-textured bread crumb. The long-grain type yields a sandy dry crumb. The former types are known to have a relatively low amylose content and a low GT, while the latter has intermediate values for these characteristics. One can therefore conclude that a low amylose content and GT are necessary for proper baking of rice flour into bread. These breads met sensory reference standards better than US long rice grains. One must also mention that bread made from wet-milled rice flour gave a superior texture than those made from dry-milled flour. However, one must remember that wet-milled rice flour is more problematical because of its higher cost and therefore dry-milled flour is more usually used. Fortunately, high-speed mixing can largely overcome some of the drawbacks associated with dry-milled flour. Use of various enzymes has also been found to produce an appropriate texture for the bread crumb. Various approaches have thus been used, and are still being investigated, for overcoming the deficiencies of rice flour in terms of yielding a soft and well-leavened bread.

Apart from bread, other rice-based baked products are also being widely made. These include various types of cakes, cookies, crackers and so on. The objective is the same, viz. to provide snack foods for wheat-sensitive population as well as for other reasons explained. In the West these productions inevitably take the form of a modern technological process. Here the processes are basically the same as used for wheat flour, except that rice flour is substituted for wheat flour and the formulae modified as required to obtain an acceptable and attractive product.

The GT of the rice plays an important role in formulation of products containing a high amount of sugar, particularly sweet cakes. The high sugar content in the formula takes up too much water and hence pushes up the GT of the starch markedly. This fact precludes the use of high- or intermediate-GT rice for baking. The reason is, in that case, the GT is raised to such a high level in absence of sufficient water that the material does not get to be properly cooked and the cake collapses. Besides, the cake gets a sandy texture. For leavened rice cakes, on the other hand, high-amylose, hard-gel and high-setback rice gives better cakes with more fluffy and flaky structure. However, it may have a higher rate of retrogradation which has to be taken care of.

Many types of crackers, cakes and cookies have been traditionally made in many Asian rice countries by traditional methods (see later). Many of these are now being made by modern industrial techniques in Japan and to varying extent also in Taiwan, Korea and possibly China. Wet-ground or partially wet-ground rice flour usually gives better products than dry-ground flour for these products. Even though more expensive, such wet-ground or partially wet-ground flours may be produced and marketed in these countries, to meet the requirement for these very popular snacks. Sometimes, some of these industries are in the stage of transition from home- to cottage- to large-scale industries. As an example of some of these snack products, one may cite the case of *senbei* and *arare*. These are two very popular crackers in Japan. These are typical of those that have been cooked as well as baked. *Senbei* is made from usual nonwaxy japonica rice but *arare* is made from waxy rice. In both cases, the flour is cooked, pounded and kneaded, made into shapes and then baked. Both products come in diverse forms, sizes and shapes. These two are only examples, being very popular in Japan, but otherwise many more types are produced throughout the rice-eating regions of southeast and east Asia. Locally available rice, which usually means low- or intermediate-amylose rice in these regions, is used for making these products. Production of these traditional crackers generally also involves a certain amount of puffing. Depending on historical situation, some of these products may now or shortly be made by modern industrial techniques.

Another well-known baked product is the unleavened flat bread popular in the Indian subcontinent and central Asia, called the *chapati* or *roti*. It is a product usually made from wheat flour, but sometimes may have to be made with other cereal flour because of the reasons cited. Rice flour does not make good *chapati* due to lack of gluten, nor can its dough be easily rolled. However, reasonable products can be made if the dough is prepared from partially cooked rice flour (or if dough is made with hot water). Rapid retrogradation during storage is another problem with these *chapatis*. Use of hydrocolloids such as guar gum, xanthan, lucust bean gum and hydroxypropyl methylcellulose improves the texture of the product and keeps it more extensible during storage. Addition of fungal alpha-amylase further improves the texture of rice chapati.

8.4   Modern granular rice snack foods

8.4.1   Puffed, shredded and flaked rice

Apart from above products made starting with rice flour, various snack foods are also made in the West starting with whole rice grain. These are puffed rice, rice flakes, shredded rice, and so on. The objective again is partly to meet the requirement of wheat-sensitive population and other reasons cited, but also to add variety to people's food. As will be discussed later, flaked rice and puffed rice are well-known products prepared in the Indian subcontinent. But those are made by traditional methods not quite suitable for modern industrial techniques. The products under discussion here on the other hand are those that are made by modern techniques in the image of well-known snack foods made traditionally from wheat or maize. Japan is in an unique position of both being a rice-eating and modern industrial society. Both types of products are therefore being made there. This is also partly true of Taiwan, Singapore, Hong Kong and perhaps China.

Puffed rice in the Indian subcontinent is made by a method of parboiling followed by heating (see later). The process followed in the West is quite different. Here milled rice is cooked with sugar syrup and steam under pressure. Non-diastatic malt syrup and other ingredients may also be added. The cooked rice is broken up and partially dried and tempered. The rice is surface dried by passing under a source of radiant heat and then lightly rolled or bumped by passing through a pair of rolls. The bumped grains are cooled, tempered and briefly toasted, when the grain gets puffed. The cooked grains may also be coated with skimmed milk, brewers' yeast, wheat germ meal, etc. to provide flavour and nutrition. This oven-puffed rice cereal is a popular rice snack. It may also be flavoured with sugar syrup and other coatings and colourings.

The other, more common, method of producing puffed rice is by the gun-puffing method. The principle here is to subject the uncooked milled rice grain within a chamber to high pressure, usually by superheated steam, and then to suddenly release the pressure by activating the gun The rice puffs out from the sudden release of pressure. The exact pressure, temperature, timing. etc. are all critical factors and are generally proprietary information. The puffed grain may be seasoned and flavoured with various syrup and nutritional adjuncts to make the product more attractive to the consumer.

Shredded rice is another popular product made from milled rice. It is prepared by cooking washed milled rice in a rotary cooker with sugar, malt, syrup and salt under pressure. The cooked rice grits are subjected to a preliminary drying in a louver oven to produce a dry and hard grain surface which allows the grain to flow freely. It is tempered and then passed though a pair of shredding rolls. One of the rolls is smooth while the other has a series of shallow, square or rectangular corrugations running around its periphery. The individual grains get caught between the rolls and are drawn into the corrugation. Thereby a lattice-like ribbon is produced. These are then passed through an oven and toasted. It is said that short-grain rice varieties produce a better product than long-grain rice. Apparently, this has nothing to do with rice's amylose content or the GT but more with the size and shape of the grain. The thin long grain tends to lie flat into the groove during rolling and hence fails to weld.

A relatively new ethnic snack food is made industrially in the USA. It is a disc-shaped puffed product, low in calories, made from whole-grain milled rice. It is made by pressing milled rice grains in a heated mould between two heavy plates. After heating, the upper plate is suddenly raised, causing a sudden release of pressure. The rice gets puffed and fused together into a disc. No added binder is needed. Minor ingredients such as sesame, millet, salt and other flavouring agents may be added.

Rice flakes are also being made, though this is not a very common product. The process is the same as in making puffed rice. Milled rice is cooked, moisture adjusted, rolled and puffed (blistered). In the case of puffed rice, the cooked grains are bumped less and puffed more. But in the case of rice flakes, the cooked grains are flaked more and then lightly blistered. The product is basically similar to corn and other cereal flakes.

8.4.2   Miscellaneous

Various products such as snack bars, crispies, rice cakes, crackers, chips and rice drinks are also manufactured and marketed in the industrialised countries. Earlier and historical versions of some of these products used to be made at home and may be still prevalent in Asian societies in some cases. Now-a-days various rice fries and other products are being made by the process of extrusion, replacing the use of baking and frying because of its versatility. Rice drinks made from brown-rice flour and dispersed with vegetable oil and emulsifier are now available in the market. These are easily fortified with vitamins and minerals and are fairly popular drinks, especially in Japan.

8.5   Traditional rice products of Asia

As we have just seen, a modern rice-products industry has lately sprung up in the West and also in Japan. This has happened during a little over the last half a century or so due to historical reasons. Although modern and sophisticated in terms of technology and enterprise, this industry is too small in terms of volume compared to the rice products produced in the world as a whole. The lion's share of these products in terms of volume and versatility is produced in the traditional rice countries of Asia. This is not surprising considering that over 90% of world's rice is grown precisely in these countries. Besides, these products are being prepared not just now but have been so produced for centuries and perhaps millennia. Historical circumstances being what they are, these traditional products are being produced largely by techniques prevailing or being evolved over millennia, initially only at homes, now perhaps being supplemented by production in cottage-scale or at best some small-scale enterprises.

As already mentioned at the beginning of this chapter, there are some curious regional specificities in terms of the types of these products. Thus the bulk of the products in southeast and east Asia are cooked cakes, pan cakes, fries, crackers, puddings and allied foods prepared from wet-ground rice. Another interesting feature is that the rice being used for this purpose is perhaps as often as not waxy rice. The situation in the other important rice region, viz. south Asia, is somewhat different. Cooked and formed cakes and puddings prepared from wet-ground rice are no doubt produced and used here as well. But these do not dominate the scene. It is three granular products – flaked, puffed and popped rice – prepared from whole-grain rice that dominate the scene here. Secondly, it is the common non-waxy rice of the region that is most widely used for preparing the products. Waxy rice is rarely used. What little use waxy rice is made of is mainly in the north-eastern corner of India which, significantly, abuts the south-eastern Asian region. There is another interesting difference between these two regions. Noodles are widely used as substitutes for cooked table rice for the main meal throughout southeast and east Asia. These are traditionally being made from wet-ground rice batter or paste, even though much of the prevailing noodles may be made from wheat flour now. In contrast, noodles are rarely used in south Asia. These regional differences in the prevalence of the product types should be a matter of considerable historical, geographical, cultural and anthropological significance.

8.5.1   Traditional products from wet-ground batters and pastes

As already explained, a variety of cooked or semi-cooked rice products are traditionally being made, starting with wet grinding of the rice, throughout the rice countries of Asia for centuries and millennia as snacks or breakfast. While this is true of all the countries of the region, it is more so especially for southeast and east Asia. In addition to products meant for breakfast and snack food, another similar product being made in this region similarly starting with wet-ground rice is noodles. Most of these products are being made usually from wet-ground rice batter or paste, prepared as part of an integrated process for making the products, and then cooked. Of late, some of these products may be sometimes made starting with dry flour produced by dry or wet or semi-wet grinding of rice. But this is rare, being especially true if at all of Japan and perhaps a few industrialising centres in Taiwan, Hong Kong or Singapore. These traditional products are now discussed below.

8.5.2   Noodles

Noodles are an interesting product. Noodles are primarily consumed as a part of the main meal and not as supplementary snack or breakfast. Interestingly, these are produced starting from whole (or broken) grains, which are first ground and then reformed into a new structure. Humans have a liking for structured food. And new structure, shape and texture provide variety and freshness. So sometimes structured grains are first pulverised and then formed into structures of different kinds to impart variety and newness to the diet. Another advantage in favour of noodles in the case of rice is that one can use broken rice that is invariably produced to a varying extent during milling of rice. Noodles, pasta, macaroni are well-known structured food products prepared from wheat flour in the West, particularly in Italy, as a new way of preparing food from a staple. In the same manner, rice can be ground and then reformed into noodles. This is a very popular food throughout southeast and east Asia.

Noodles are made from wet-ground rice. The paste is formed into a dough and steamed or otherwise partially cooked in water. It is then kneaded and finally extruded in a simple press either manually at home or mechanically in the cottage industry. The extruded noodles are cooked either by steam or by boiling in water and then dried. There are many variations of these processes both for tiny or small industries. Dry-milled rice flour can also no doubt be used for preparing noodles. But this is not usually done. Partial surface gelatinisation occurs in the flour upon dry grinding of rice, whereby the particles become somewhat soft and sticky. Besides dry-ground flour does not have a uniform particle size. As a result, some particles remain coarse and they produce a rough structure.

Rice quality plays an important role in noodle quality. Low-amylose rice varieties do not make good noodles. High-amylose varieties are needed to provide a more ordered structure and proper strength. The high-amylose, low-GT rice such as Taichung Native 1 variety produces the best noodle structure. Interestingly, the Japanese, who never use indica rice for their meal, yet import indica rice for preparing rice noodles. Often moong legume (*phaseolus aureas*) starch is added to rice flour for noodle making to improve the amylose content and hence the noodle strength.

Recently it has been found that presoaking of the rice in sulphurous acid, as is done in the case of maize for preparing corn starch, is helpful in noodle preparation. The sulphur dioxide partly dissolves the protein and thereby loosens the grain structure and partially frees the starch granules. This helps in the subsequent steps and produces a superior product.

8.5.3   Rice cakes and fries

Numerous kinds of cooked and semi-cooked rice cakes have been traditionally made throughout the rice countries of Asia. Well known among these are different kinds of *mochi* in Japan, *puto* and *suman* in the Philippines and *idli* in India (Fig. 8.4). These are being traditionally made at home for centuries and millennia. These are generally made from fully or partially cooked wet-ground rice, rarely from rice flour. Sometimes these are also being made from cooked whole rice grain pounded into a dough. The dough prepared either from cooked batter or cooked grain is kneaded and then finally cooked in water or by steam. The cooked dough thus obtained may be either wrapped in various leaves and boiled or steamed (*suman* in Philippines). Or the cooked dough may be cut into pieces and then toasted and seasoned (*mochi*) or, alternately, again wrapped in leaves and steamed or boiled and fried (*mochi*).

Or sometimes the cooked dough may be fermented (*puto*). These are only some of the popular illustrative cases mentioned, but there are many other kinds.

There are often specific requirements of this or that rice for specific products known in traditional practice. Some of the factors behind such requirements are known but some are not. Some of these products are specifically made from waxy rice. In some cases, it has been found that waxy rice with a specific GT range is alone suitable for the product. Some *mochis* or *sumans* have specific need of waxy rice. Waxy rice having a higher GT has usually been found to produce a poor product. Puto of Philippines is best made from aged intermediate-amylose rice.

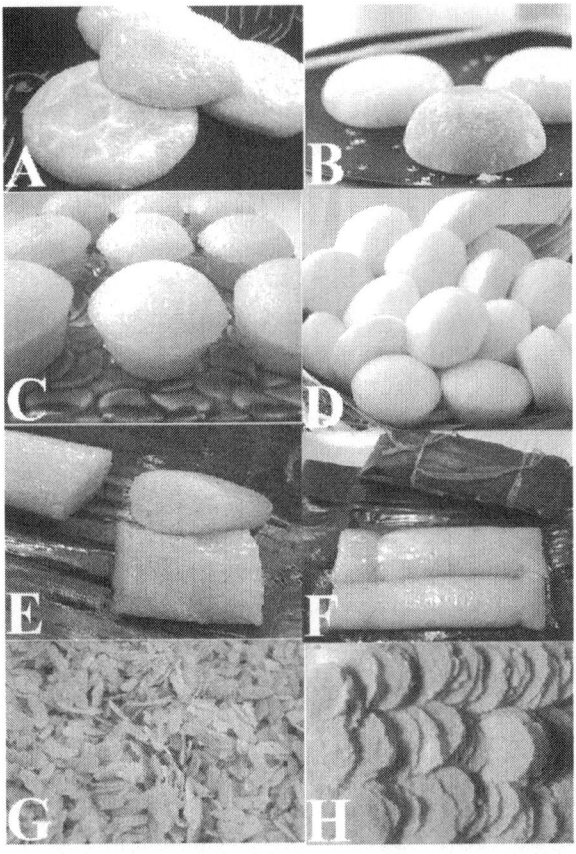

Figure 8.4 Photographs of *mochi* (top row), *puto* (second row), *Suman* (third row), *pinipig* and *emping* (last row) [Reproduced from Bhattacharya (2011) where original sources are indicated]

8.5.4   Fermented cakes

The best known fermented rice cake is the idli (Fig. 8.5). It originated in south India but has lately become very popular throughout the country. Rice, as mentioned before, does not have a network-producing protein, such as gluten of wheat. Therefore it cannot be leavened by itself. Black gram (*phaseolus mungo*) is therefore mixed with it to enable the product to rise. The Legume contains a surface-active protein and a viscosity-stabilising polysaccharide. These two together form a network film which enables the gas to be retained and thus the product to rise. Approximately two parts of rice and one part of decorticated black gram is used. The rice used is usually parboiled rice and ground relatively coarsely. The black gram is separately ground very fine. Both are wet ground. The mixed batter is allowed to ferment overnight generating gases which are retained by the network mentioned. The fermented batter is finally cooked in steam in the form of a cake. *Dosa* (Fig. 8.5) is made similarly but it is like a pancake fried in oil and not steamed.

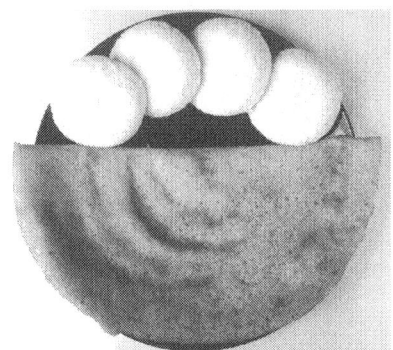

Figure 8.5 Photograph of *idli* (top) and *dosa* (bottom)

Idli and *dosa* are traditional foods of south India, where the prevalent rice varieties have a high amylose content. Indeed high-amylose rice has been found to be most suitable for making the products. Reasonably acceptable *idli* could be made with intermediate-amylose rice, but low-amylose rice gives unacceptable *idli*.

There are other fermented rice cakes. *Puto* is one. Some *puto* may be made directly but usually it is made after fermentation. Interestingly, *puto* being a product of Philippines, where prevalent rice have intermediate amylose content, best puto is indeed made from intermediate-amylose rice. *Laochao* is a Chinese fermented sweet rice cake. Milled waxy rice is fermented with fungi and yeast. *Laochao* is made from such fermented steamed waxy rice.

8.5.5   Miscellaneous

Puddings and sweets are also being regularly made from rice in the rice countries of Asia. Rice pudding is more often than not made from waxy rice by boiling in milk with or without the addition of egg yolk, sugar, vanilla and fruits. Many variations are prevalent. Many sweets and desserts are made, generally from waxy rice, but also from non-waxy rice traditionally in the rice countries of Asia.

8.6   Traditional products made from whole-grain rice

A majority of rice snack products made in southeast and east Asia are cooked or steamed or fried cakes, puddings and crackers made from wet-ground rice. But the situation in south Asia is somewhat different. Here, the three most popular rice products are different as a class. They are dry products made from indirectly cooked whole-grain rice – viz. flaked rice, puffed rice and popped rice (Fig. 8.6). No doubt many cakes and fries made from wet-ground rice are prevalent in this region also. The most important among them are the popular *idli* and *dosa* made in south India, now become popular throughout the country. However, the three granular dry products mentioned above are by far the most widely produced and consumed rice products in the region as a whole.

Popped rice   Puffed rice   Flaked rice
                   (Expanded rice)

Figure 8.6 Three common Indian rice products

One should note that, as already mentioned, many rice products in the West too are no doubt made by puffing or frying or flaking of cooked whole grains, with or without the addition of malt, sugar and other adjuncts. But those are quite different products. First, those are made starting with milled rice cooked in water. Second, those are produced using modern sophisticated technology in the image of other cereal snacks and breakfast foods made in the West. The products being referred to in south Asia, on the other hand, are produced by an integrated traditional process starting with paddy.

8.6.1   Flaked rice

Flaked rice or beaten rice is probably the most popular whole-grain rice product made in the Indian subcontinent. This food used to be made at home or at best in a cottage-scale industry. One should note that this flaked rice is quite different from the rice flakes made and marketed in the West. The latter is quite similar to other cereal flakes such as corn flakes that are consumed as a breakfast cereal along with cold or warm milk.

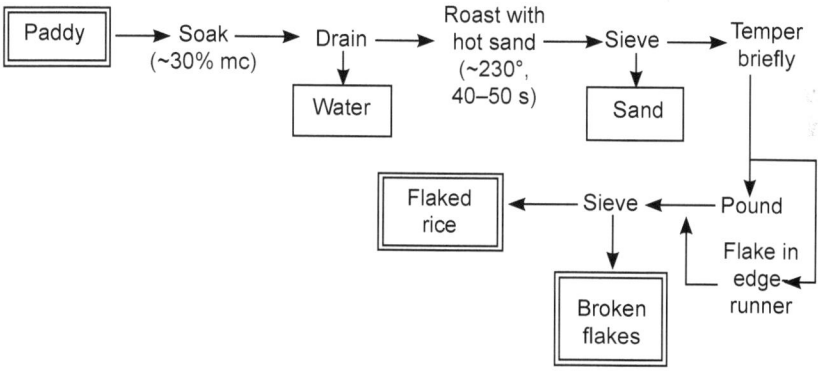

Figure 8.7 Traditional process for making flaked rice. mc = moisture content (wet basis) [Reproduced with permission from Bhattacharya (2011)]

Flaked rice in India is a simple but versatile product made by a simple technique developed through experience over centuries. The underlying principle is that paddy is parboiled by dry heat (as now understood) as a part of an integrated total process. The intermediate dry-heat parboiled grain is then immediately flaked by pounding or rolling (pressing) (Fig. 8.7). Paddy is soaked overnight in warm water and roasted with hot sand in small batches (1–2 kg) in an iron pan over a strong fire. The hot roasted paddy – which, unknown to the producers, is now actually a dry-heat parboiled paddy – is immediately flaked. In olden days, when it was made at home, the flaking was done by pounding the paddy in a wooden mortar and pastel. The husk and outer bran of the grain become very dry and friable upon roasting, while the inner starchy rice grain becomes plastic. The husk and the bran therefore get powdered and easily removed by pounding, while the plastic inner grain

becomes flattened. When lately manufacture of flaked rice became a cottage-
or small-scale industry, the pounding in the mortar was replaced by flaking in
an edge-runner. The latter is a wide, circular, horizontal rotating sieve pan on
which a small idle roller is mounted at the edge (Fig. 8.8). Hot roasted paddy
is put on the rotating pan, when the grains get repeatedly pressed between
the raised edge of the pan and the idle roller. As a result, the friable husk and
the outer bran get powdered and pass out through the sieve. The plastic inner
starchy grain gets flattened or flaked and remain on the pan. The thickness of
the flake is adjusted by adjusting the time of pressing in the edge-runner.

Figure 8.8 Photograph of edge-runner [*Courtesy:* S.S. Engineering Works, Kolkata]

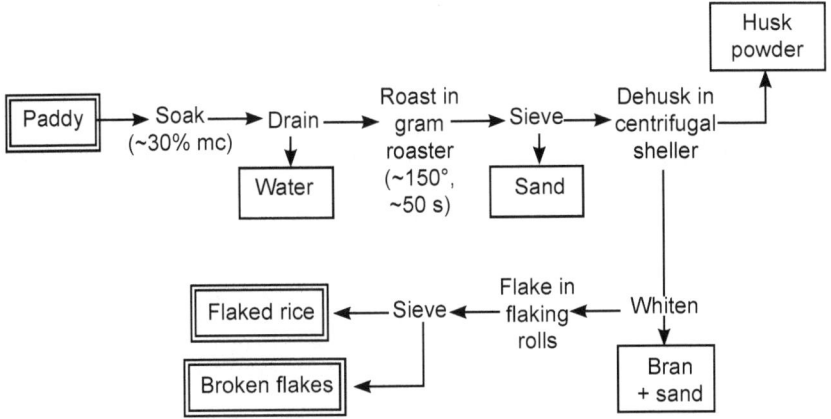

Figure 8.9 Improved process for making flaked rice. mc = moisture content (wb)
[Reproduced with permission from Bhattacharya (2011)]

The above process has recently been further improved. The batch roasting of a small quantity of soaked paddy in an iron pan over a fire is replaced in this improved process by roasting in a continuous 'gram roaster' (Fig. 7.14). The continuously produced roasted paddy is then flaked in several edge-runners. Alternatively, a process that has been demonstrated but is yet to be commercially adopted is as follows. The roasted paddy produced continuously from the gram roaster is immediately dehusked continuously in a centrifugal sheller, then aspirated and lightly whitened (debranned) in a polisher to remove some bran (Fig. 8.9). The lightly milled grain is then immediately and continuously flaked by a pair of heavy-duty flaking rolls. The thickness of the flake can be adjusted by adjusting the clearance between the rolls.

Flaked rice is an extremely versatile product. First, being a dry and compact product, it can be stored, carried to places and traded. Second, it is a precooked product and virtually ready to eat. Third, its water absorption capacity is very high. This is because of many factors. One, being a dry-heat parboiled product, its hydration capacity is already high. Two, the increased surface area caused by flattening also helps in hydration. Three, the flaking too causes further starch damage and increases the hydration capacity. As a matter of fact, flakes of appropriate thinness can get 'cooked' even in cold water. It can thus be considered almost as a ready-made or disaster meal and has been known to have been carried by travellers in olden days, along with pieces of jaggery (unrefined cane sugar), while travelling to long distances. Third, soaked flaked rice can also be eaten with warm milk as a ready breakfast. Fourth, being a precooked starchy product, it can be puffed by heating by virtue of being gelatinised. Finally, it can thus also be made into a variety of sweets or savouries with or without precooking, seasoning, sweetening, or frying. Or it can be puffed by toasting or heating in oil and consumed with or without further sweetening or seasoning.

A more or less similar product is very popular in the Philippines but is made from waxy rice. It is called *pinipig*. The best *pinipig* is made from Malagkit Sungsong variety, which gives the best water absorption capacity as well as stickiness. Low-amylose rice or even high-GT waxy rice does not produce good *pinipig*. Another similar product is *emping* in Malaysia (Fig. 8.4).

Another traditional flaked and puffed rice product is *hurum* in the Assam state in the northeast of India. It is made by dry-heat parboiling of selected waxy (*bora*) rice varieties. The soaked and roasted paddy is first pounded to obtain an intermediate flaked rice as above. The flakes are then rubbed with fat and puffed by roasting.

There are two other broadly similar products made in the same Assam state and surrounding areas. One is *Bhaja chawal* and the other is *Sandahguri*.

Both are a kind of dry-heat parboiled rice and then lightly puffed. The former is made from waxy (*bora*) varieties but the latter from low-amylose (*chowkua*) paddy. Both are eaten as snacks with milk or otherwise sweetened.

Yukwa is a traditional oil-puffed Korean snack food made from waxy rice. It is made similarly by soaking the waxy rice in water followed by dry-heating or boiling. The resulting intermediate parboiled rice is puffed by toasting. The hardness and expansion ratio of the product are important quality parameters and are affected by processing conditions as well as variety.

8.6.2   Puffed rice

This is another popular traditional snack food in the Indian subcontinent. As already indicated above, puffed rice is made outside the subcontinent also. But those processes are different. In the Indian process, paddy is first parboiled and milled, after which the milled grain is puffed by sudden heat. The whole is an integrated process. Parboiling, milling and puffing are performed sequentially as part of single integrated process. Interestingly, initially it was not even known that parboiling was involved. The matter became clear only when researchers studied the system. In contrast to this process, puffed rice is made in the West without parboiling. There raw milled rice is puffed by one of two methods. Either the rice is cooked (somewhat akin to parboiling) and then is oven puffed. Or else the raw milled rice is directly puffed by gun-puffing. Here again the involvement of parboiling in the oven-puffing process was realised only when researchers got interested in studying the process.

Three factors are involved in best puffing in the Indian system. One is the variety. All varieties do not puff equally well. The amylose content of the variety plays a crucial role. Secondly, parboiling is another important factor. Raw milled rice cannot be puffed by high-temperature, short-time (HTST) heating. It can be puffed only by gun-puffing. On the other hand, once the rice is parboiled, i.e. gelatinised, any gelatinised starchy material including rice gets puffed when subjected to HTST heating. The manner and the degree of parboiling also therefore play a part in the efficiency of puffing. The third is the state of the grain at the time of puffing and the actual conditions of HTST heating. These three factors are discussed below.

The variety (i.e. amylose content) and the parboiling conditions are mutually interrelated. Parboiling as we know can be done either by steam method or by dry-heat method (see Chapter 7). In the steam method itself, again, the steaming can be done under different pressures. It has been found that for each amylose content in the rice (i.e. the variety), there is an optimum steam pressure which gives the highest degree of puffing (i.e. the highest expansion ratio). For example, for atmospheric pressure steaming (i.e. at 0 kg/sq cm, gauge), waxy rice (no amylose) gives the best puffing. But for higher amylose contents, steaming under higher pressures gives still better expansion. Overall, highest expansion is obtained from varieties having 27.5% amylose at an optimum steaming pressure of 2.0 kg/sq cm (gauge) This is for steam parboiling. But a still better puffing is obtained if the paddy is dry-heat parboiled and further better when the paddy is 'pressure parboiled' (i.e. low-moisture, high-pressure steam parboiling). Thus the variety (i.e. amylose content) and the parboiling conditions are mutually interrelated and best puffing expansion is obtained under optimum conditions.

As for the state of the rice grain and the precise HTST conditions, these are best explained in the process diagram of the most widely adopted cottage-scale process (Fig. 8.10). Paddy is soaked overnight in warm water. The water is drained and the paddy is roasted in hot sand in an iron pan with stirring over an open fire for a short time. The sieved product is dry-heat parboiled paddy. This is then milled usually in a local Engelberg huller. The resulting milled dry-heat parboiled rice is the material which is finally puffed. It is first preheated down to about 10.5% moisture, lightly brined (salt solution) and then puffed by HTST heating. Finally sand is removed by sieving. Many variations of the process exist. In some parts of the country, the rice is parboiled by boiling in water and then milled and puffed. As mentioned, it was not even realised that parboiling was involved in the processes before researchers went to study them. The application of brine just before puffing improves the expansion ratio to some extent but mainly helps to produce a smooth surface of the puffed product. Cracked grains do not puff well.

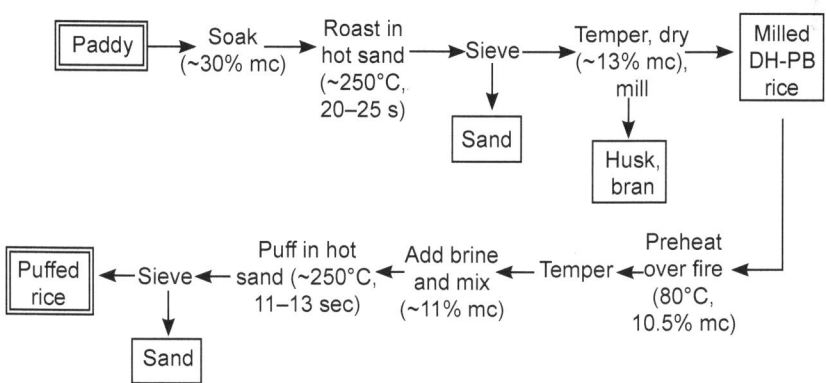

Figure 8.10 Process for puffed rice. mc = moisture content (wet basis); DH-PB = dry heat parboiled rice [Reproduced with permission from Bhattacharya (2011)]

Gun-puffing of raw milled rice is also prevalent in parts of southeast Asia as a small-scale or cottage-scale industry. Raw milled rice is taken in a small, iron, cylindrical receptacle. The cylinder is heated to a high temperature over a fire, thereby increasing the pressure inside. This pressure is suddenly released by actuating a catch, when the grains puff and come flying out from the cylinder to be caught in a net. One unfortunate drawback of gun-puffed rice is the blackening of the scutellum part of the milled rice (the point at which the germ is attached to the endosperm). This black spot in the puffed rice often creates consumer doubt and resistance. It will be noted that in this process, it is a case of ungelatanised starch being puffed. For this reason it requires the building of a high pressure and then its sudden release. In the traditional Indian system, the starch is first gelatanised and so gets puffed at a lower temperature without a sudden trigger.

Puffed rice, like flaked rice mentioned earlier, is another popular Indian snack food. It is widely used for breakfast or snack after seasoning or sweetening. Its only drawback is its large volume which makes its storage and transport rather difficult.

8.6.3   Popped rice

This is a product largely specific to the Indian subcontinent. It is also a fairly popular snack food in the region, although not as much as the other two products mentioned. Here paddy is directly puffed without any further processing just by HTST heating of suitably pretreated paddy in hot sand. The product and the process are similar to those of popcorn.

Two factors are involved in ideal popping: Viz., the variety and the popping conditions. All varieties do not pop well. Proper selection of the variety is therefore important. It has been found that the varietal specificity is related to the structural factors of the paddy grain. Popping requires the building up of a high pressure inside the paddy grain and then its sudden release (somewhat like gun puffing). So the grain structure must be strong enough to retain a high pressure built by heating which generates steam and hot air within the paddy grain. Finally, a point comes when the structure is no longer able to contain the high pressure, so the pressure is suddenly released and the inner starchy grain pops out like a flower. The structural features involved are three-fold. One is the degree of husk interlocking. The more tight the interlocking, better the retention of pressure and the popping expansion. Second is the hardness of the endosperm. The harder the grain, the better the pressure retention and hence popping. Finally, another factor is an endosperm defect in the form of 'white belly' (chalky area) on the ventral side (germ side) of the rice grain. This chalky area at the germ side forms a weak spot which causes premature leaking of pressure. White belly is therefore detrimental to popping. Secondly, optimum grain conditions and HTST conditions are also necessary for best popping. These are best explained by the process flow sheet (Fig. 8.11). Interestingly, it has been found that the paddy should first be dried to about 9% moisture before readjusting to 14% moisture. The former (9% moisture) is probably required for optimum tightening of the husk preparatory to heating. The latter (14% moisture) may be ideal for generating the correct steam pressure. HTST heating is provided by quick heating with hot sand.

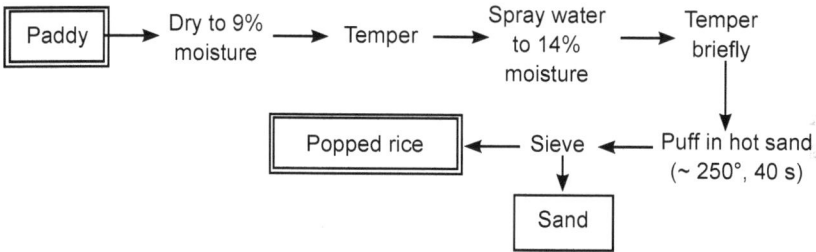

Figure 8.11 Process for popped rice [Reproduced with permission from Bhattacharya (2011)]

Popped rice is also a fairly popular snack and breakfast food in the Indian subcontinent. Its advantages and disadvantages are more or less similar to those of puffed rice. It is a ready snack or breakfast, and can be eaten with a variety of seasoning or sweetening or with milk. But, like in the case of puffed rice, its large volume is a handicap for storage and transport.

8.7   Miscellaneous industrial products

Rice is also being used to make some unusual, mostly non-food products, often with the main purpose of utilising the available broken rice. Some amount of broken rice is invariably produced during the milling of rice, some special use for which has had to be found in industrialised countries. Preparation of pet food and use in brewing have thus been developed as the major outlets of broken rice grain, especially for the very small broken pieces (less than half grain length). As a matter of fact, uses for these two purposes constitute the main outlet for rice (actually broken rice) other than for cooked table rice in the West.

8.7.1   Pet food

Pet foods are the latest industrial products made from rice. It may possibly be the largest industrial outlet for rice. Rice has some special advantage in respect of production of pet food. These are its easy digestibility and low allergic response. Rice is therefore preferred for producing pet food especially wherever the comparative price is favourable. Brewers' rice (tiny broken pieces) is most commonly used as input for obvious reasons. However, rice flour, bran or even whole-grain rice can be equally well used, if circumstances require. Extrusion process is the most common method of processing.

8.7.2   Alcoholic drinks

The other important industrial outlet for rice is to use it for producing alcoholic drinks.

In the case of Japan, *sake* is something special. It is their national drink. It is not made as a means of utilising broken rice. *Sake* is made because it is wanted. And it is made not from broken rice but from specific whole rice which is specially milled. A high level of scientific enquiries has gone into various aspects of *sake* making for a long time, especially during the last half a century or so. Today it is a highly science-based and modern industry in Japan. Not only the fermenting organisms, the enzymes involved, the precise conditions of fermentation, even the degree to which the rice has to be milled for *sake*-making is a matter of rigorous scientific scrutiny.

Traditional alcoholic drinks from rice are made at home and cottage-scale industries throughout the rice countries of Asia also. But these are not national drinks. And little science has gone into making of these products till now. Traditional knowledge and rules of thumb are what guide their production.

Fine broken pieces of rice (brewers' rice) are widely used as an adjunct in the modern brewing industry in the West. The most important requirement for such use is the rice's GT. A low GT is helpful in conserving energy for cooking as well as in keeping the temperature sufficiently low such that the enzymes are not inactivated. Equally or more important is the uniformity of GT of all the grits. A uniform GT is helpful in that the entire batch can get gelatinised simultaneously. A low protein content is desirable.

8.7.3   Germinated brown rice

Germinated brown rice (GBR) is a recently developed rice product being produced in Japan and Korea. When brown rice is soaked in water under suitable conditions, germination is triggered. Simultaneously many chemicals and micro-nutrients start being produced. Some of these micro-nutrients are said to have great nutritional benefits for humans. It has been observed that Ɣ-aminobutyric acid (GABA) increases dramatically if brown rice is soaked at 40°C in water for 8–24 hours. GABA is said to have very great beneficial effects on human health. Inositols, ferulic acid, phytic acid, trocotrienols, oryzanol, etc. are also said to be produced during germination. All these nutrients have beneficial effects for consumers. It is said that GABA reduces blood pressure, helps in ameliorating sleeplessness and is also helpful in menopausal and presenile conditions. A good deal of interest in this nutrient has recently been generated in Japan and it is a subject of intense research. Another favourable point of GBR is, unlike untreated brown rice, the germinated grain cooks in the normal manner. GBR is now available as a commercial product in Japan.

Further reading

Details of various rice products and the methods of their production or manufacture are given in various text books of rice. The more important among the text books are :

- Houston (1972)
- Luh (1980)
- Juliano (1985, 2007)
- Luh (1991 b)
- Champagne (2004)
- Bhattacharya (2011)

9.1 Byproducts of rice milling

The rice grain that is obtained after harvesting and threshing of the crop from the field is in the form of paddy or rough rice. This paddy comes clothed in the protective cover of a pair of glumes, called the lemma and the palea. The two halves together go by the name of the husk or the hull. It forms roughly one-fifth (20–22%) of the weight of the clean paddy grain. This husk is woody, siliceous and abrasive and is therefore inedible. It has to be removed before the rice grain can be eaten.

The inner grain obtained after the removal of the husk is called the brown rice (or dehusked or dehulled rice). The brown rice has some outer layers, namely the pericarp, testa (or tegmen or seed coat), the nucellus and the aleurone layer (see Fig. 9.1). These layers are fibrous and somewhat impermeable and make the grain rather difficult to cook. It is therefore necessary to remove at least a part of these layers also to make the grain edible. The fact that cultural changes and commercial interests have tended to get these outer layers removed in full by the rice-milling industry before the grain is offered to the consumer, is a different matter. In any case the fact is that a part or the entire portion of outer layers of the brown rice grain is also removed by the milling process before preparing the rice for consumption.

This act of removal of the two sets of outer layers of rice is called by the generic name of milling of rice. The process of removal of the husk alone is known as shelling or dehusking or dehulling and the act of removal of the outer layers of the brown rice is called as milling, polishing, whitening, pearling or simply debranning. The two processes together are also known by the common generic name of rice milling. The brownish powder that comes out during the process of whitening of the brown rice is known by the commercial name of rice bran. Unlike the woody and siliceous husk, which is unfit to be used even as an animal feed, the bran is potentially edible and is a good animal feed.

The amount of bran removed during whitening is called the degree of milling, i.e. the percentage weight of the brown rice that has been removed during its whitening. Thus if 5%, say, of the weight of brown rice is removed during the process of whitening, one can say that the rice has undergone 5% degree of milling. This degree of milling varies from market to market or culture to culture. To obtain fully white and shining rice, it is necessary to remove 10% or more of the weight of brown rice as bran (10% degree of milling). Indeed in modern industrial societies and high-value markets, the milling imparted to rice is routinely of 10% degree or more. But in rural areas or in traditional markets, the degree of milling is usually less and can be even as low as 3–4%.

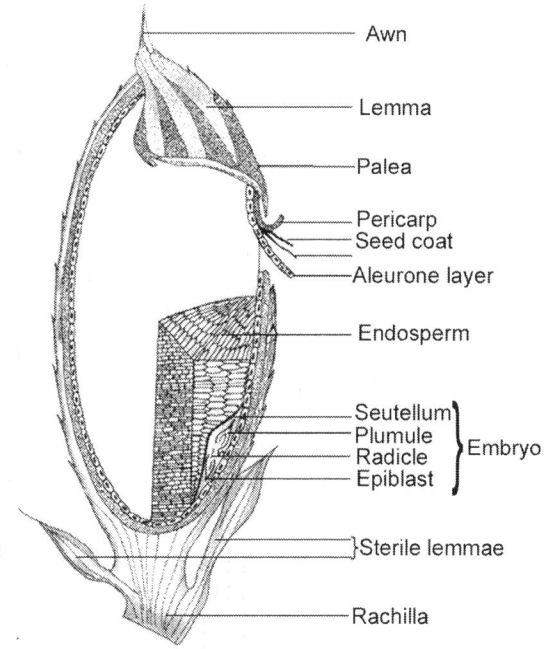

Figure 9.1 Line drawing of a cut section of a paddy grain

We can thus say that 100 units (by weight) of clean paddy produces roughly 20–22 units of husk, 3–8 units of bran and 70–75 units of milled (or white or polished) whole and broken rice grains after the process of milling of rice. The husk and bran are thus the two byproducts of rice milling. For the health of any industry, utilisation of its byproducts is very important. Fortunately both husk and the bran are valuable materials and have many uses. Utilisation of husk is discussed in the next chapter. The properties and uses of bran are discussed here.

9.2   Composition and properties of rice bran

9.2.1   Types of bran

Rice milling (discussed in Chapter 5) is not a uniform or fixed process. It comes in different systems. The amount and the nature of the bran thus vary from system to system.

In full-scale rice milling, the husk is removed first by shelling (dehusking). Shelling itself can be carried out in two different systems. One is by the under-runner (or disc) sheller and the other by the rubber-roll sheller (see Chapter 5). In the under-runner sheller, the dehusked grains have to travel some distance over the face of the lower disc covered with emery. Therefore, a tiny but finite amount of bran is scratched out during this travel. In addition, a part of the husk gets powdered. Therefore, a small but finite amount of powdery material is recovered by sieving the under-runner sheller output. This powdery material is referred to as the sheller bran. Thus this material too goes by the name of 'bran'. Although this 'bran' does contain a very small amount of true or real bran, as explained, the latter's proportion is very low. So the value of this sheller 'bran' is nil or negligible. In the case of rubber-roll shelling, hardly any powdered husk is produced nor does any scratching of the outer layer of the brown rice occur. So it hardly produces any 'bran'.

Rice is sometimes milled in huller machines (Chapter 5). This was a common practice in early twentieth century. The dehusking and whitening operations are combined in a single step in this system. Therefore, the real bran that is produced here is greatly diluted by a large amount of powdered husk. As the powdered husk is woody, siliceous and inedible, its admixture makes the value of this total hullar bran very low. So it is not much worth any effort at its utilisation.

The rice bran of industrial importance is the true bran, i.e., to be precise, the outer layers of brown rice separated during the process of whitening of the brown rice. Here again, two types of whitening machines are in use (see Chapter 5). One is the abrasion-type (emery type) and the other the friction-type (metal type) machine. In abrasion-type whiteners, the bran is cut out in small pieces from the surface of the brown rice by the sharp particles of emery. As a result, the bran comes out in small particulate pieces. Also the random abrasion causes some endosperm (inner grain) matter also to be cut and get mixed with the bran. This bran thus contains a slightly lower amount of oil and other contents. In friction-type polishers, on the other hand, bran is removed by peeling by the mutual rubbing of the grains under pressure. The bran therefore comes out in slightly larger pieces and with little or no endosperm parts. This bran thus contains slightly more oil and other useful constituents as compared to the bran from abrasive polishers.

Bran composition changes slightly with the stage and extent of whitening as well. When the dehusked brown rice is whitened in more than one stage, the contents of oil, fibre and ash of the bran gradually decrease while that of starch gradually increases from the first to the last whitening stage. However, these differences are not important in real-life practice. The bran coming out from successive whiteners are invariably mixed together in practice and handled together.

Bran composition is affected by the degree of milling also due to the same reason as above. The contents of oil, ash and fibre gradually decrease and that of starch gradually increases as one goes from the outer-most to the inner-most layer of the brown rice. As a result, the contents of these former constituents gradually decrease and that of starch gradually increases as the degree of milling is increased, especially beyond 4–5%.

The presence or absence of germ in the bran is another reason why its composition can differ slightly. The germ is attached to one corner of the rice grain (as shown in Fig. 9.1). This germ gets gradually scraped out or separated during whitening of dehusked rice. The pieces then get mixed with the bran. Rice bran obtained from rice mills therefore includes rice germ (about 20% by weight). In some European (Spain, Italy) milling systems, the germ is got separated from the bran by sieving and aspiration. But the germ is not separated and remains in the bran in other countries. As the composition of germ is slightly different from that of the powdered outer layers of the brown rice, the composition of the resulting bran differs somewhat accordingly.

Another reason why the composition of the bran can vary somewhat is pretreatment of the paddy, especially parboiling. As discussed in Chapter 7, the grain oil contained in the oil bodies located in the aleurone layer tends to move out into the soft outermost layers of the grain during the process of parboiling. As a result, oil is more in parboiled-rice bran than in raw-rice bran. Besides, the bran tends to come out in larger peels or flakes from the parboiled grain.

Finally, in certain European and US milling systems, the rice is first whitened in whitening machines as usual. But after that the thus whitened grains are subjected to what is called 'polishing' operation in certain special equipments. In these 'polishers', the abraded grains are rubbed by leather straps instead of being subjected to further abrasion or friction. The purpose is to remove all loose bran particles and a small part of the endosperm so as to give the rice a good shine and polish. The material thus removed is called 'polish' as opposed to 'bran' that is obtained from the whitening process. The composition of this 'polish' is therefore slightly different from that of the bran. The polish contains less oil, protein, fibre and ash as compared to the bran but contains more starch. But this practice is hardly in vogue in other countries especially in Asia, where often the act of whitening itself is referred to as 'polishing'. So the terms 'polish', 'polisher' and 'polishing' are to be used with care and should be understood in the context of the discussion.

9.2.2   Composition of rice bran

As is clear from the above discussion, composition of the bran is not a definite, invariable entity. It not only varies perhaps from variety to variety, but also according to the mill type, the procedure of milling adopted, the type of whitening machine used, the stage and extent of milling, and any preprocessing of paddy. However, taking a broad view, rice bran may be considered to have an average composition as shown in Table 9.1. Compositions of organic material, i.e. material derived from living systems, are generally given in terms of protein, lipid (fat or oil), carbohydrates (nitrogen-free extract), minerals (generally expressed as ash), crude fibre (derived from certain groups of undigestable carbohydrates) and microconstituents (such as vitamins, etc.). This is called the 'proximate composition'. It is in terms of such proximate composition that the data in Table 9.1 are expressed.

Table 9.1 Proximate composition of rice bran and germ

Constituent	Unit	Range of values in	
		Bran	Germ
Protein	%	12–17	15–25
Fat (oil)	%	13–23	17–40
Carbohydrates	%	34–54	35–42
Starch	%	10–25	2–4
Free sugars	%	3–7	8–12
Pentosans	%	5–8	5–7
Hemicellulose	%	9–17	9–15
Cellulose	%	5–10	3–7
Lignin	%	3–10	1–5
Crude Fibre	%	6–14	2–10
Ash	%	7–15	5–11
Phosphorous	%	1–3	1–3
Potassium	%	1–2.5	1–2
Magnesium	%	0.5–1.3	0.4–1.5
Silica	%	0.5–1.5	–
Calcium	mg/100 gm	30–130	20–100
Iron	-do-	9–50	10–50

Contd...

Contd...

Constituent	Unit	Range of values in	
		Bran	Germ
Vitamins			
B$_1$ (Thiamine)	-do-	1.2–3.0	4–8
B$_2$ (Riboflavin)	-do-	0.2–0.4	0.2–0.5
Nicotinic acid	-do-	25–50	2–10
E (Tocopherol)	-do-	2.5–13	8–10

The above composition and its background can be better understood if one refers to the internal structure of the rice grain, and the trend of variation of its composition, across the grain's cross-section. The cross-sectional structure of the rice grain was shown in Fig. 9.1. Now the bran would clearly correspond to the external layers of the brown-rice grain, viz. the pericarp, the tegmen, the nucellus, the aleurone, possibly the subaleurone and perhaps a little bit of the endosperm of the rice grain. While composition of these individual botanical tissues may not be easily available, a rough idea of the matter can be obtained from another source. Dr. Salvador Barber of the Instituto de Agroquimica y Technologia de Alimentos (IATA) (Institute of Agricultural Chemistry and Food Technology) in Valencia, Spain and his colleagues meticulously collected the layer-wise composition of the brown rice grain from the outer skin to the grain centre. These results were summarised in a figure as shown in Fig. 9.2. This figure shows two things. First it shows that the rice grain does not have a uniform composition throughout the grain. Actually, the composition varies as per the specific location within the grain, presumably corresponding to the botanical tissues as shown in Fig. 9.1. Specifically, the data show that the amount or concentration of the various classes of compounds (protein, fat, ash, etc.) changes as one moves from the outermost layer to the centre of the brown-rice grain. Most components have their highest concentration just or slightly inside the outermost layer and then gradually decline towards the centre of the grain. The only exception is starch, whose concentration progressively increases from the outermost to the innermost layer. This figure gives one a good idea of the expected composition of the rice bran. Clearly the bran would contain much or most of the rice-grain oil (fat), more amount of protein as compared to the original grain and most of the proportion of fibre and minerals (ash). Starch would be proportionately much less and would increase as the degree of milling increases (as one approaches and abrades the inner grain or endosperm). This is precisely what can be seen in the average composition of rice bran as shown in Table 9.1.

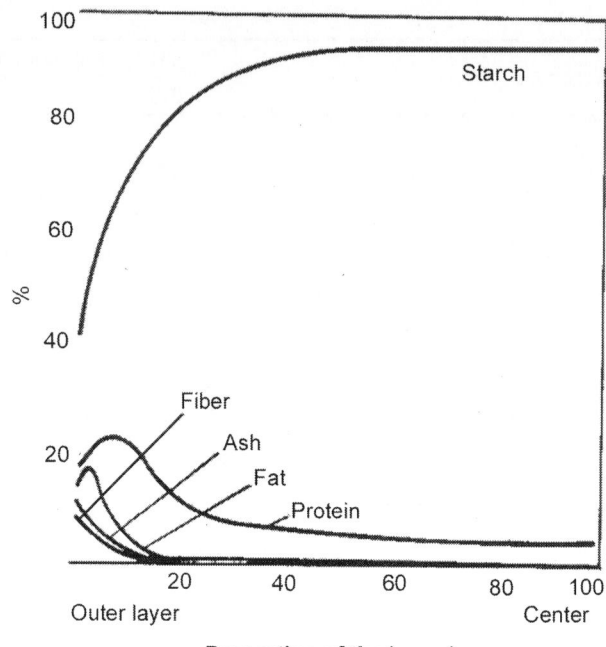

Figure 9.2 Distribution of major milled-rice constituents within the brown rice grain [Reproduced with permission from Barber (1972)].

Carbohydrates constitute some 30–35% of the bran. These consist of varying amounts of starch (depending on the degree of milling, i.e. how much of the endosperm has been abraded). The rest of the carbohydrates constitute the cell wall or similar materials which together constitute the crude fibre, consisting of cellulose, hemicellulose, lignin, pentosans, etc. These constitute the so-called fibre of the bran which is indigestible. This fibre is both a desirable and an undesirable part of the bran as will be discussed below.

The protein content in rice bran is much more than that of the original grain. However, this is only a part of the good story. The quality of the protein is also superior. Proteins in living systems can be classified into four groups depending on the solvents in which they are soluble. The four classes of proteins and their solubility pattern are as follows:

- Albumin – soluble in water
- Globulin – soluble in salt solution
- Prolamin – soluble in alcohol
- Glutelin – insoluble in most solvents (soluble in alkali)

The highest amount of protein in the whole rice grain is glutelin (largely located in the endosperm). Proteins are giant compounds, composed of several amino acids joined together. (Amino acids are about twenty in number, nine of which are essential for us and must be obtained from our food intake. The rest are nonessential because these can be made in our body from other amino acids). Lysine is one of the essential amino acids. Glutelin, which forms the bulk of the rice's endosperm protein, is relatively deficient in lysine. Much of the small amounts of albumin and globulin of the rice grain are located in the outer layers. These two classes of protein are relatively rich in lysine. Bran protein therefore contains more lysine and hence is nutritionally superior to the protein in the milled rice.

Many of these bran proteins are actually enzymes (enzymes are what make chemical reactions go on within the living cell). As the rice grain is a living seed, it contains many enzyme systems to carry out numerous reactions during its germination. Most of these enzymes are located within the germ and aleurone layer. These enzymes are largely composed of albumins and globulins. That is how rice bran is richer in albumin and globulin as compared to the original rice grain. As will be discussed later, some of these enzymes are important in terms of utilisation of rice bran.

Rice bran contains relatively large amounts of minerals as can be seen from its high ash content in Table 9.1. Predominant among the minerals is phosphorus. It amounts to up to 3% in bran. It is a component of a molecule called phytin which is rather high in rice bran and is one of the negative factors in terms of utilisation of bran. Another negative factor is silicon, part of which may come from sand as a contaminant of bran. Calcium and iron, the two important nutrients, are rather low.

Most of the B vitamins in rice grain are located in the outer layers of the grain including the germ. Most of these vitamins therefore come into the bran during the milling process (rendering the milled rice very poor in these vitamins). Rice bran is therefore rich in B vitamins.

Rice bran contains a relatively large amount of oil or fat. This is a matter of much economic and industrial importance and will be discussed in detail in a separate section.

Parboiling is an important premilling processing of rice. South Asia in particular is the home of rice parboiling. About 65% of rice in south Asia is parboiled. India is the largest producer and consumer of parboiled rice where a little over 50% of its paddy production is parboiled. Little parboiling is done in countries outside south Asia (see Chapter 7). This aspect is mentioned here, for parboiling has a significant effect on the composition of bran. On the one hand, as explained in Chapter 7, parboiling has the effect of disrupting the oil globules located in the aleurone layer, whereby the oil is pushed outwards into the testa, pericap, etc. As a result, the oil content of the bran jumps from about 15% in the case of raw rice to about 20% in that of parboiled rice. Secondly, as again explained in Chapter 7, parboiling has the effect of firmly retaining some of the micronutrients, especially B vitamins and minerals, in the endosperm. Consequently, these micronutrients get firmly attached to the milled rice and are not removed into the bran during milling of the brown rice. As a result, conversely, parboiled-rice bran contains much less B vitamins and minerals as compared to raw-rice bran.

The composition of rice germ is broadly similar to that of bran (Table 9.1). But rice germ contains substantially more protein and fat (oil) and less fibre than rice bran. It also contains less starch and more free sugars. Here again, most of the protein is albumin and globulin and germ contain slightly more lysine than in bran. Like bran, rice germ also contains high amounts of phosphorus. Here again, about 75% of phosphorus is in the form of phytin. Vitamins and other minerals are again high in germ, but germ contains more thiamine and less nicotinic acid as compared to bran.

9.2.3   Properties of rice bran

As the nature and composition of the bran itself vary according to circumstances, as explained in the above sections, the properties of the bran also would naturally vary accordingly.

The particle size distribution in rice bran varies depending on its source. Friction-type whiteners produce bran of a rather larger particle size as compared to abrasion-type whiteners. The former tends to peel out the bran by mutual friction between the grains, while the latter cuts out small particles of bran by the sharp edges of the emery. Not only that, in abrasion-type whitening, the particles tend to become finer as the milling stage increases from the first to the last. Again, heat stabilisation (discussed later) tends to bring out agglomeration of the bran particles. Similarly bran from parboiled rice is flakier with a larger particle size than bran from raw rice. The pieces of germ in the rice bran are generally bigger in size than the other components. Dr. Barber and his group in IATA, Valencia, Spain did extensive study of the histology and histochemistry of the various particles in different brans.

The bulk density (i.e. the weight per unit volume of bran in bulk) is very low (see Table 9.2). But the true density (i.e. the weight per volume of the bran particle as such, excluding the air space entrapped between the particles), is close to that of the milled-rice grain. Its angle of repose (angle formed with the horizontal plane of the ground by a heap of the material) is about the same as that of rice and paddy. Its equilibrium moisture content (i.e. moisture content attained when exposed to air) is about 2 percentage points less than that of milled rice. In other words, the highest safe moisture content for its storage (moisture content attained when exposed to an atmosphere of about 70% relative humidity) is about 12%.

Table 9.2 Some properties of rice bran

Property	Bran	Germ	Milled rice
Bulk density (kg/litre)	0.2–0.4	0.5	0.8
True density (g/millilitre)	1.2–1.3	–	1.45
Angle of repose (degree)	38	–	38
Equilibrium moisture at 75% RH (% wet basis)	12	–	14

9.3   Potential of rice bran as a source of oil

9.3.1   Genesis of rice bran becoming a source of oil

The brown-rice grain contains only about 2.5–3% fat or lipid (i.e. oil). The rice grain should thus be of no consideration as a potential source of commercial vegetable oil. However, four factors have combined to make the rice grain as a good source of vegetable oil. First, Nature has so devised that the bulk of the grain oil is located in an ultrathin layer within the outer parts of the brown rice (aleurone) including the germ (this is true not only of rice but of all food grains). Second, one should note that rice is milled as a whole grain, not as flour or powder (as is the case of wheat, maize, sorghum, etc.). Third, the inner grain (endosperm, more or less equivalent to the milled grain) is harder than the bran tissue. As a result of these three factors, a soft concentrated oil-bearing layer is more or less neatly scraped out of the hard inner grain during whitening or milling of rice. A fourth factor is germ. In other cereal grains, the germ is bigger in size and gets separated as a whole during milling. In rice, the germ is smaller and gets mixed with the bran. This adds to the oil content, for the germ is richer in oil. The net result of all this is that the bulk of the grain oil comes out greatly concentrated in the powdery bran. That is why the rice grain, apparently of no importance as such as a source of oil, becomes a significant source of oil in the form of its oil-enriched bran (12–20% oil).

Parboiling also influences the oil content of bran. When paddy is parboiled, the oil bodies in the aleurone layer get disrupted and the oil content is pushed out into the outermost layer. As a result, more of the grain oil comes into the bran from parboiled rice. Parboiled-rice bran thus contains no less than about

20% and even up to 25–30% oil. South Asia (the Indian subcontinent), which accounts for roughly one-third of the world's production of paddy, is the home of rice parboiling. About 65% of the paddy produced here is parboiled. One can therefore say that the average oil content of rice bran in south Asia would be not less than about 18–20%.

This is the genesis of the unlikely rice grain, through rice bran, becoming an important source of vegetable oil for human consumption. Calculated at a modest rate of 6% bran on paddy (8% bran on brown rice) and 15% oil in bran, the potential availability of oil from rice bran is no less than 5.5 million tonnes of vegetable oil from rice in the world (2000–2010). However, there are many impediments to the full realisation of rice bran for being utilised for production of oil. The impediments are primarily three-fold: (i) problems of mobilisation of bran, (ii) availability of other sources of vegetable oils in the country, and (iii) instability of the oil in bran.

9.3.2   Impediments to the use of rice bran as a source of oil

The first impediment is the mobilisation of the bran. The 'Rice Country' (see Chapter 2) where 90% of world's rice is grown and consumed is, by and large, even now (2015) in the stage of developing and is economically and technologically still lagging behind. Much of the rice crop is thus milled in hundreds of thousands of tiny to small to medium-size rice mills scattered all over the vast area of south, southeast and east Asia. Mobilisation of the entire amount of the bran, potentially huge in aggregate, is thus a mirage. Besides, much of the rice crop is still a subsistence crop in these areas, produced by poor farmers for their self-consumption. This paddy, perhaps amounting to 25–50% of the total production of the country, is milled by custom milling in tiny amounts. This bran therefore exists only on paper and can never be mobilised for any industrial use (no doubt being utilised for feeding domestic livestock and poultry). It is thus only the fraction of the rice that is milled in the relatively large and modern milling systems that can be potentially available for production of rice-bran oil. This fraction may at best be considered as of now to be somewhere within the range of 20–50% of the total rice production in these countries.

But here again, the urge for utilisation of bran as an oil source would depend on the second factor, viz. the availability of other sources of vegetable oil in the respective country. Getting the oil out of rice bran is technologically more difficult than that from oilseeds etc. So bran processing would have a low priority if the country is endowed with other oil sources in abundance.

9.3.3   Instability of the oil in rice bran

There is also a third problem, viz. the problem of oil instability. The oil present in rice is meant to supply energy and nutrition to the growing plant as the seed germinates. So Nature has provided an enzyme in the seed to break down the oil for further utilisation at the appropriate time. Oil is a compound formed by combination of glycerol and fatty acids. The job of the enzyme is to break the oil down (hydrolyse) back to glycerol and free fatty acids (FFA). These compounds are then utilised by the growing seed. In the intact seed, the oil is mostly present in the aleurone layer. But the enzyme, called lipase, is present in the testa layer. So they do not come in contact until the stage of germination. However, during the process of rice milling, the two get unfortunately mixed together and the lipase action starts. The oil starts getting broken down to FFA very fast. This FFA, if it is too much in the oil (say more than 4–8%), makes the oil unfit for edible purpose. Oil with high FFA is difficult to refine also because of too high refining loss. This is one of the hazards of industrial production of oil from rice bran. In other words, solving the problem of lipase is a priority if the full potential of utilising rice bran as a source of oil is to be achieved.

There are three ways to address this issue. One way is to extract the oil as soon as, or soon after, the rice is milled and the bran is produced. So the lipase does not get much time to hydrolyse the oil. This approach requires the existence of a well-organised system of quick bran collection and oil extraction. Alternatively, in very high-capacity rice mills, as for example that exist in Japan as explained below, one may have an in-house extraction system producing oil from the bran almost simultaneously as it is being produced.

It appears that this approach of quick collection and oil extraction has been successfully organised not only in Japan but fairly successfully in India as well. The problem is virtually nonexistent in Japan due to their milling system, as will be explained shortly. While most rice mills in India are nowhere being as large as in Japan or the West, sustained efforts by various agencies over the decades seem to have succeeded in developing an efficient and well-organised system of quick collection of bran and its rapid extraction. This is proved by the figures of rice-bran oil (RBO) production (see Table 9.3 below), the lion's share of which is being marketed as edible vegetable oil.

Where quick collection and immediate oil extraction cannot be organised, two alternative approaches are possible. One is to produce the oil (RBO) nevertheless and to use it only for industrial purpose. The high FFA would prevent the oil from being made edible, but the oil can still be used for industrial purposes, such as for manufacture of soap. The other alternative is to so treat the bran as to destroy the lipase soon after the bran is produced.

Once the lipase is destroyed, the bran can be collected and extracted at leisure. This is the classical approach of bran stabilisation which will be discussed in a subsequent section. Meanwhile, one may note in this connection that this approach of bran stabilisation for the purpose of production of edible-grade RBO, although strongly promoted for decades from 1970s onwards, does not seem to being followed anywhere in the world. However, be it noted that stabilisation is being promoted again now from another angle. This will be discussed later below (Section 9.5).

Table 9.3 Approximate production of rice-bran oil in the world (around year 2000)

Country	Thousand metric tonnes
Bangladesh	1.5
Brazil	1.5
Cambodia	4.6
China	90
India	472.7
Indonesia	0.15
Japan	65
Korea	11.7
Republic of Korea	9.2
Laos	2.6
Myanmar	17.6
Nepal	7.6
Pakistan	3.7
Sri Lanka	5.5
Thailand	7.8
Vietnam	7.6
Total	722.2

Reproduced, with permission, from Orthoefer and Eastman (2004), who sourced the data from FAOSTAT, 1998.

9.3.4   Status of production of oil from rice bran in the world

A rough idea of the production of rice-bran oil (RBO) in different countries in the world is shown in Table 9.3. It can be seen that the highest production of RBO in the world happens to be in India, the next distant highest production being in Myanmar (Burma), China and Japan. But these values do not give a true picture of the extent of utilisation.

Production of RBO as a fraction of the total potential in the country is highest in Japan. About 85% of the available rice bran is being utilised there for production of oil. This is because of several factors. First, the rice-milling system is somewhat peculiar in Japan. Shelling or dehusking of paddy crop there is carried out by individual farmers or groups of farmers in the farm itself. Only dehusked brown rice is collected by the centralised rice mills for whitening. These rice mills are therefore huge in capacity and a large quantity of brown rice is being handled by each mill. Rice bran is for this reason available in Japan in huge quantities from individual rice mills in centrally located regions. Collection and mobilisation of rice bran for processing is thus not a problem at all in that country. Secondly, this ready collection also avoids the problem of instability of the bran oil discussed above. Thirdly, Japan is deficient in production of other oilseed crops. Utilisation of rice bran as a source to obtain vegetable oil therefore comes naturally to Japan.

The utilisation proportion of rice bran as a source of vegetable oil is also very good in India. It is about 65% of the potential. Considering the current socio-economic status of India, and the prevailing structure of the rice-milling industry therein, this high proportion is most remarkable. This achievement seems to have come about because of two or three factors. One, India has been chronically short of oilseed production for a long time. Much of her vegetable oil need has had therefore to be imported. There was therefore always a strong incentive to utilise all potential sources of oil. Secondly, there has been a very active Solvent Extractors' Association of India (SEAI). The Association has been strenuously promoting and lobbying for the utilisation of rice bran as a source of oil for many decades. Third, there has always been a strong governmental and institutional research support in this field. As a result of all these factors, the production of RBO has been steadily going up for many decades as shown in the approximate figures below:

- 1970–71 – 20 thousand tonnes
- 1980–81 – 130 thousand tonnes
- 1990–91 – 200 thousand tonnes
- 2000–01 – 600 thousand tonnes
- 2010–11 – 800 thousand tonnes
- As of today – 900 thousand tonnes

In comparison, the production of RBO in Myanmar (40% of potential), China (6%), Thailand (25%), though fair in quantity, is not very appreciable in terms of the potential. Production elsewhere in other countries in Asia is negligible, presumably primarily because these countries are endowed with other sources of vegetable oil (oil palm, coconut, soya, oilseeds).

9.4   Production and use of rice-bran oil

9.4.1   Extraction, refining and use of rice-ban oil

As the oil content of rice bran is rather low, oil cannot be easily pressed out from it as from oilseeds. Pressing was a common practice in Japan in olden times (before World War II) and has also been practised in China, Vietnam and Thailand. The process involves steam cooking of bran under pressure, followed by drying and hydraulic or expeller pressing under very high pressure (up to 1000 kg/cm2). A little over half the bran oil can be pressed out by these methods. One advantage is that the crude oil is of purer quality than that obtained by solvent extraction. However, this method of bran oil preparation has been now given up in Japan and also elsewhere in favour of extraction by solvents, mostly hexane.

Solvent extraction can recover almost the entire amount of oil in the bran (residual oil in bran, 1–1.5%). But it has the disadvantage of extracting many impurities and colour so that refining of the oil is more expensive. The methods largely follow the conventional extraction and refining technologies, with modifications as needed to suit the powdery material and other issues.

Extraction can be of batch, battery or continuous type. The batch and battery systems are generally operated in Japan. Continuous extraction plants have larger capacity and are being operated in many places (India, Myanmar, Egypt, Mexico, Taiwan and Thailand).

Batch extraction is in individual extractors. The miscella (mixture of oil and hexane solvent) is passed through a filter to an evaporator for desolventising. In battery extraction, several extractors are arranged in series, and miscella obtained from one is passed through the other in a counter-current system. Fresh solvent is applied only to the last extractor. This is a semi-continuous system.

Continuous extraction achieves the highest economy in steam, power, labour and material. It is suitable only when the capacity is at least 50 tonnes of bran per day. It uses the counter-current principle. While in the batch or battery system bran as such can generally be used, its pelletisation is essential in the continuous system. This is necessary to reduce the problem of channelling and of fines which clog the filters as well as pass through the filters to make the oil turbid. To prevent this, bran is made into pellets of 6–8 mm size. Stabilisation by moist heating (see below) causes agglomeration in bran in which case pelletisation may not be required. Stabilisation of bran by extrusion cooking produces flakes, in which case pelletisation is unnecessary. In batch plants, where pelletisation is not essential, some processors add coarse husk to help extraction; the husk is later on removed from the dry deoiled bran by sieving.

Although hexane is the solvent of choice and has been universally used, other solvents may have some special advantages. Ethyl or isopropyl alcohol have been tried. These have the advantage of extracting less wax, which means dewaxing problem is eliminated, and of extracting sugars and B vitamins which can be recovered as a byproduct syrup. However, alcohol extracts more colour, causing difficulty in bleaching, and reduces the nutritive value of the extracted bran by dissolving the B vitamins.

Parboiled-rice bran contains more oil and should therefore be a preferred material for extraction. However, processors have often reported problems with it. First, parboiled bran is more difficult to pelletise and may require some special conditions. To overcome this problem, some processors mix raw bran with parboiled bran which, however, may cause FFA problem. Second, predrying of parboiled bran is essential for batch extraction. Third, parboiled bran needs a longer time and more number of extractions to extract all its oil. In batch extractors parboiled bran needs at least three solvent washings to recover all oil. Finally, improper parboiling may darken the oil colour.

9.4.2   Refining of rice-bran oil

Refining is a must if rice-bran oil is to be used as a cooking oil. RBO is being used for cooking in Japan for a fairly long time. In India its use as a cooking oil has started only recently in the last few years. Until recently RBO (high FFA) was being used in India either for industrial purpose (mainly soap) only. Or, low-FFA (8–10% FFA) RBO, designated as 'edible grade', was being directly used for making plastic hydrogenated fat. In this, the pooled oil was being deacidified, bleached and deodorised before hydrogenation. The wax was not being removed, for the wax got diluted in the pooled mix.

Other vegetable oils are often prepared by pressing and can be directly used in food. But rice-bran oil is prepared by solvent extraction and as such it has to be refined for use in cooking. Refining is also necessary in view of the presence of acids in it. Even if the oil is prepared from fresh or stablilised bran, the amount of FFA in the oil is usually more than 3%, making it generally unsuitable for direct use in food.

Refining of RBO is more difficult than other vegetable oils due to three reasons: (1) Alkali refining of bran oil causes much higher loss of oil than in other oils due to reasons not fully understood. (2) Rice oil, unlike other oils, contains wax, removal of which is a little complicated. (3) Heat treatment, including improper parboiling and improper heat stabilisation, and especially in presence of high FFA, causes darkening and fixation of colour in the oil, which is difficult to remove by bleaching.

Refining of crude RBO involves the following steps as described below.

Dewaxing

Rice-bran oil is unique in possessing a fairly high amount of wax. The wax causes much difficulty in subsequent refining and therefore must be removed before other steps of refining are attempted. The wax is soluble in oil-hexane miscella at usual extraction temperatures (60–70°C) but is virtually insoluble at low temperatures. A double extraction system is therefore one approach to dewaxing, although this is expensive. The bran is first extracted with hexane at a low temperature (4–5°C), when the wax remains insoluble, giving a wax-free oil. The bran is then extracted a second time at 60°C, which on cooling yields wax as a precipitate. Alternatively, the conventional hot-solvent extraction followed by cooling of the miscella or of the desolventised oil can separate wax by deposition. The separated wax can be collected by centrifugation or filtration or by slow settling. However, as the viscosity (thickness) of the oil increases on cooling, separation of wax from oil by cooling is time consuming and hence expensive in tropical countries. Dewaxing in miscella (mixed with hexane) is easier as the viscosity is thereby reduced. But the process involves additional cost. Recently it has been observed that separation from the oil can be more easily achieved by adding a solution of a surface-active agent which wets the wax particles and helps their merging and separation.

Degumming

Gums and mucilages in vegetable oils generally consist of phosphatides and glycolipids. They may also include sugar and protein complexes in colloidal form. Prior degumming of oil is important, for polar lipids have surface-active properties and form an emulsion with the soap-stock formed during deacidification. So the soap-stock does not easily separate and occludes a lot of oil, leading to high losses. Phosphatides, if recovered, are valuable byproducts and can be used like lecithin. In case of deacidification by distillation or during heating for other reasons, prior degumming is a must. Otherwise high discolouration occurs during heating which is very difficult to bleach. Degumming is normally carried out by adding small quantities of phosphoric or citric acid followed by filtration or settling. Alternatively, steam can be passed into the oil up to a temperature of 80–100°C, or the oil can be heated with addition of a little water, which separates gum by flocculation (mutual adhesion and floating). Recently two new techniques of degumming have been tested in India. In one, calcium chloride is used as a reactant for simultaneous degumming and dewaxing. This is being widely used in Indian industry now. In another method, phosphorylase is used for enzymatic degumming.

Deacidification

Removal of FFA from rice bran oil generally causes much problem. Due to some unknown reasons, high losses of oil occur during its alkali neutralisation. However, prior dewaxing and degumming followed by carefully controlled addition of caustic soda and controlled time and temperature of heating can reduce refining losses to a minimum. One reason of the high loss of neutral oil during alkali refining is the thin and fluffy nature of the soap-stock. It does not separate easily and therefore occludes considerable amounts of oil. Reducing refining losses requires hardening of the soap-stock to make it settle easily. Certain chemicals such as sodium silicate, sugars (molasses), ethanolamine and some alcohols and glycols seem to achieve this. Soap is obtained as a byproduct of alkali neutralisation. As the oil is mostly unsaturated, the soap is soluble in cold water.

Miscella refining is an effective method of reducing refining loss, for the soap-stock separates easily in this case. Refining in the miscella has the additional advantage of giving less coloured oil. Miscella refining using binary solvents (hexane and alcohol) is also done. The soap dissolves in the alcohol layer. Soap is recovered from the alcohol layer as fatty acids. Refining loss is low, but the process is expensive.

Another method of deacidification is to remove FFA by steam distillation. In this case the free fatty acids are obtained as valuable byproducts. The oil is heated to 200–240°C under a high vacuum of 3–4 millimetres of mercury and is subjected to steam stripping. The free fatty acids distil over and are collected as a distillate. There is no refining loss. However, the process is somewhat expensive. It's another disadvantage is that the oil colour tends to get deepened and fixed, leading to bleaching difficulty.

It has been suggested that high-FFA oil can be converted to low-FFA oil by any of the physical refining systems (distillation, miscella refining), and the low-FFA oil can then be further refined by alkali treatment or used for hydrogenation.

Bleaching, deodorisation and winterisation

Bleaching of rice-bran oil is carried out by conventional bleaching earths and activated carbon. Most pigments can be readily absorbed by bleaching earth or destroyed by heat. But oxidation, particularly in the presence of traces of iron or copper, can lead to fixing of colour and resistance to bleaching. Heating in presence of gums can also lead to darkening and difficulties in bleaching. Deodorisation is carried out by heating the oil with dry steam to 200–250°C under high vacuum. It removes all volatile materials giving smell.

If the oil is to be stored at refrigerated temperatures, such as for salad oil, winterisation is necessary. The purpose of this step is to remove saturated glycerides which cause cloudiness to the oil when stored in cold. Winterisation is achieved by chilling the oil and allowing the saturated fats to settle out.

9.4.3   Use of rice-bran oil

Characteristics of rice-bran oil are broadly similar to those of other vegetable oils. Its fatty acid composition is roughly as :

Oleic acid	–	40–50%
Linoleic acid	–	30–35%
Palmitic acid	–	15–20%
Others	–	5–10%

High-FFA rice-bran oil can only be used for various industrial purposes, including for making soap. As the oil is soft (it contains predominantly unsaturated fatty acids, as shown above), it is often hardened by hydrogenation before being put to industrial use.

Refined RBO is a good cooking oil. If properly winterised, it can be used as salad oil. It has low linolenic acid and high tocopherol contents, thus rendering it more resistant to oxidation than many other vegetable oils. At the same time, high amounts of oleic and linoleic acids make it a rich source of essential fatty acids and highly nutritious. It has a good cholesterol-lowering effect. The latter effect is caused, apart from the unsaturated fatty acids, primarily by the oryzanol and other tocols etc. in it. Oryzanol, tocopherols, other tocols, ferulic acid, wax, other unsaponifiables, etc. in the oil have powerful physiological and antioxidant properties. Their presence thus renders the oil highly nutritious and to have significant cholesterol-lowering, anti-ageing, anti-diabetic, anti-dementia properties, apart from being hypoallergenic.

9.4.4   Byproducts of refining

Rice-bran oil contains many important materials which may be obtained during its refining (see Table 9.4). The wax is of high melting point (75–80° C) which is not found in other vegetable oils. It is very similar to carnauba wax and has many industrial and cosmetic uses. Vitamin E (tocopherols) and squalene have medicinal properties. Oryzanol is an ester of ferulic acid and triterpenoid alcohols and has powerful physiological effects and valuable pharmaceutical properties.

The above and many other chemicals and pharmaceuticals, including B vitamins and phytin, can be prepared from rice bran or the oil.

Table 9.4 Some valuable materials present in rice bran or its oil

Material	Amount (% in oil)	Use
Wax	2–6	In polish, food wraps, cosmetics, leather, etc.
Vitamin E	0.1	Medicinal, as antioxidant
Squalene	0.3–0.4	Medicinal
Oryzanol	1–2	Medicinal, as antioxidant
Fatty acids	Variable	Industrial
Soap	Variable	Industrial
Gums (phosphatides)	1–3	Like lecithin, emulsifying, wetting, dispersing
B Vitamins	In bran	Medicinal
Phytin	In bran	Food processing industry
Inositol	From phytin	Medicinal
Microchemicals	In bran	Industrial and medicinal

9.5   Stabilisation of rice bran

Rice bran contains many valuable constituents (oil, protein, vitamins, essential minerals, and nutraceuticals) and also some harmful components (enzymes, microorganisms, insects, harmful compounds). Ideally, inactivation of the harmful agents and preservation of the valuable components is the ultimate aim of bran stabilisation (as stated by Dr. Barber). However, in practice, the main aim of bran stabilisation as normally understood, is destruction of the enzyme lipase to prevent development of FFA.

The grain lipase is not the only source of FFA production in rice bran. There are several millions of microorganisms in every gram of rice and bran and possibly a large number of insects and their eggs as well. All these produce their own lipase if conditions are favourable for their growth, especially if the bran is allowed to have or absorb high moisture (> 12%). In other words, it is not only the original grain lipase but also the microorganism and insects developing in the bran that may cause deterioration of its oil. So these too have to be destroyed. Certain anti-nutritional compounds present in rice bran also may need to be inactivated. Fortunately appropriate heating can inactivate all these agents efficiently and simultaneously. Lipase in that sense in a good index. It is fairly heat stable. So conditions that are appropriate for destruction of lipase are also generally appropriate for inactivation of most of these other undesirable factors.

Stabilisation of bran to prevent FFA formation can be approached in two different ways: by drying and by heat treatment.

9.5.1   Stabilisation by drying

First is by drying. The action of lipase on oil is a process of hydrolysis (hydro = water; lysis = to break down). That is, it breaks down the oil with the help of water. So the presence of sufficient moisture in the bran is essential for the enzyme to act. Clearly if the bran is dried down to a low moisture, lipase cannot act even if it is not destroyed. For example, a system of pneumatic flash drying of bran (drying with a strong stream of hot air which dries and also carries away the bran) to a moisture content of 3–5% has been developed. Other systems of dry-heat stabilisation have been proposed. For example, heating bran in an open pan stabilises it by drying. Such methods are especially suitable for small-scale operation. But it needs a long heating time and the heating is often non-uniform.

One should again remember that these open heating methods do not actually inactivate the lipase. They indirectly prevent FFA development only due to the bran getting dried as a result of the heating. So an important limitation of this system is that the treated bran must be stored such that it remains dry. If the treated bran is allowed to absorb moisture, the lipase again becomes active. The treated bran cannot be left open or even stored in jute sacks. It must be put in closed plastic-lined sacks. One should also note that other undesirable materials in bran (microorganism, anti-nutritional factors) are not necessarily affected by this drying treatment.

9.5.2   Statbilisation by heat treatment

Second, the bran lipase can be destroyed by heat. All enzymes can be destroyed by sufficient heat, and lipase is no exception. Other deleterious agents (other enzymes, microorganisms, insects, many anti-nutritional compounds) are also simultaneously destroyed thereby. Destruction of lipase by heat is related to the temperature and time of heating and the moisture content of the bran. The greater the bran moisture, the lower are the temperature and time needed to inactivate lipase. On the other hand, at low moisture contents, complete inactivation requires heating at very high temperatures for a long time. So moist heat is most effective to destroy lipase.

There are two approaches : (i) heating with addition of moisture, normally by steam, and (ii) closed heating, that is, heating the bran indirectly from outside in a closed container, in which the bran moisture is retained. Both are effective. Direct steaming inactivates the enzyme very effectively in a short time. Heating at 100°C for 2–3 minutes is sufficient. But in this case the bran absorbs some moisture and therefore it has to be subsequently dried before storage. Closed heating needs heating the bran at 105–110°C for 5–10 minutes for complete stabilisation.

Many systems of moist heating of rice bran were developed especially in India. There was considerable talk of stabilisation in India during the 1970s and 80s. Many systems were accordingly developed in different laboratories. They either involved added-moisture heating (generally with steam) or closed heating. Many of these were successful in achieving the destruction of lipase.

For example, in one process, 250 kg bran was taken in a closed, rotating, steel cylindrical vessel, the whole being heated by steam through a jacket. Continuous systems were also developed. However, as already mentioned earlier, the idea of stabilisation ultimately did not for some reason find favour with the solvent extractors. Probably the question of logistics and that of quick verification and proof of an effective stabilisation of every consignment were the major stumbling factors. On the other hand, extractors over the years have developed a system of quick mobilisation of untreated bran and its extraction into a highly effective system in India. So the question of stabilisation was neatly bypassed in India, yet utilising the bran almost up to the fullest potential for producing oil.

Another system of bran stabilisation using frictional heat for heating the bran was developed in USA during the 1980s – 90s. It involves extrusion cooking of the bran. The intense friction generates sufficient heating in the bran to stabilise it. Besides, the treated bran comes out in flakes which also helps in solvent extraction. This system appears to be highly efficient but is capital intensive. It did not find favour in India because of the cost, where in any case, as mentioned, the question of stabilisation was bypassed. It may have been used to a limited extent in USA, Japan and Korea.

However, we should note that this system of stabilisation by frictional heat appears now to being revived for another reason. Strong efforts have recently been on for some time for use of bran as human food (see below). It is in this effort that this renewed emphasis on stabilisation is appearing.

While on this subject, one should note that parboiling per se is an efficient method of producing stabilised bran. Parboiling involves soaking of the grain in water followed by its steaming. Clearly the grain lipase is thereby completely destroyed. However, experience of solvent extractors with parboiled-rice bran has not been good in India. Actually the reason lies in the prevailing practices. Rice millers often refrain from drying the parboiled paddy fully. The resulting somewhat high-moisture bran thereby cannot escape microbial and insect contamination with its obvious consequences.

9.6   Use of rice bran as food and feed

9.6.1   Nutritional value of rice bran

As already discussed, rice bran contains valuable nutrients which make it an important component of animal feed and potentially usable in human food as well.

The protein content of bran is around 15%,. This clearly puts rice bran as an important source of protein. The protein goes up further in deoiled bran (17-20%). The amino acid composition of rice bran protein compares favourably with those of other cereal brans, oilseeds and even cereal grains including wheat and maize. The chemical score of the protein as per the FAO/WHO reference pattern is 80% for first limiting amino acid level (lysine) and 90% for the second limiting amino acid level (threonine). Another method of measuring the nutritive value of a protein is by feeding it to young rats and measuring the 'protein efficiency ratio' (PER), that is, the gain in the rat's weight in gram per gram of protein eaten. The PER of rice bran protein is around 1.8–1.9 ; that of germ protein is slightly higher (2.0). These values compare favourably with those of grains like wheat (1.0), maize (1.2), groundnut (1.7) and rice (1.9). Cow's milk has a PER of 2.5 and egg 3.8. However, the digestibility of bran protein (in the rat) is not very good. This is caused by the high amounts of fibre and phytin in bran and by the fact that most of the protein is in the form of protein bodies.

Digestibility by sheep (non-ruminant) of bran organic matter (protein, lipids and carbohydrates) is good, but it is low in the case of fibre. Defatting improves digestibility. Hence, the final nutritional value as well as the calorific value of the defatted bran is nearly the same as that of original full-fat bran (for sheep). Presence of husk in bran reduces digestibility. Digestibility of hulls for non-ruminants is practically zero, although some hulls are customarily fed to poultry. The amount of total digestible nutrients in bran is higher than that in wheat bran and is quite comparable to wheat endosperm and oats. It needs hardly to be said that digestibility of rice bran in ruminants (cow, buffalo) is obviously excellent.

Full-fat bran is obviously a good source of calories as well as essential fatty acids.

Rice bran is an excellent source of B vitamins and vitamin E. Contents of these vitamins are reduced in bran from parboiled rice because the vitamins are partly retained in the milled rice when parboiled.

The minerals in bran and their role in nutrition are discussed below.

9.6.2   Anti-nutritional factors in bran

Rice bran contains some anti-nutritional factors, although most of these are only of minor significance.

Phytin is the most important factor. Most part (90%) of bran phosphorus exists in the form of phytin. It is located in the protein bodies of the aleurone. The phosphate groups of phytin are liable to get complexed with calcium, iron and zinc, and also with protein. These three minerals are already low in bran. Complexing with phytin further tends to render them unavailable for nutrition. Rice bran contains the highest amount of phytin among all cereal brans. When incorporated as fibre at 6% level in chick diet, only rice bran among various brans reduces growth and bone deposition.

Calcium : phosphorus and phytin : zinc ratios are important in nutrition. A ratio of 1 : 2 to 2 : 1 of the former is desirable but it is less than 1 : 10 in rice bran (but the ratio is equally low in other grain as well). Phytin: zinc molar ratio of over 10-15 depresses growth in rats. But this molar ratio is around 30–40 in rice bran (also in other grains brans). On the other hand, increasing the calcium content in bran by adding calcium carbonate (for example to help in polishing of brown rice) depresses bioavailability of zinc.

Trypsin inhibitor is a toxic factor present in rice bran in small quantities. It is an albumin and is present mostly in the germ. Hemagglutinin is another toxic factor and is a globulin. A similar factor is lectin which is a glycoprotein. But all these are present only in micro quantities and are easily destroyed by heat.

Dietary fibre, which is high in bran, can also act as an anti-nutritional factor in excessive amounts. The amount of dietary fibre as determined by enzymatic and biological methods is about four times that of the chemically determined crude fibre. Excessive fibre decreases digestibility of nitrogen and may also bind some minerals like calcium and iron.

9.6.3   Use of rice bran as animal feed

Use of rice bran in composite feeds is an obvious choice. But it has the following limitations : (1) high amounts of fibre, (2) high amounts of unsaturated fat, (3) presence of calcium and phosphorus in undesirable proportions, (4) high amounts of phytin, (5) sometimes too much sand and silica, (6) variable composition, and (7) high instability.

The dry matter of ration for an animal with a single stomach (non-ruminants) should not contain more than 6% fibre, but the fibre content of feed for ruminants may be as high as 35%. For this reason, bran cannot be used in unlimited quantities in non-ruminants.

The high unsaturated oil content of bran prevents its use in unlimited proportion in animal rations. The body fat, for instance in pigs, is thereby affected and becomes too soft. Deoiled bran naturally does not cause this problem.

The problem caused by the calcium : phosphorus ratio and phytin have already been discussed. Similarly the amount of sand and silica in compound feeds should not be more than 4%, but rice bran that has been made or handled improperly often has more.

In industrial manufacture of balanced feed, too much variation in the composition of a raw material, as is common in rice bran, is quite undesirable.

Chemical instability of bran is another factor. Fat hydrolysis develops FFA, while oxidation may lead to undesirable compounds as well as bitter taste. Deoiling of bran leads to greater stability. Heat stabilisation is the best choice. It prevents FFA development and also helps in reducing the activity of microorganisms and other pathogenic and/or toxic factors in rice bran.

Rice bran is used satisfactorily in pig diets up to 20–40% in the ration. Upton 40–50% bran can be used in the diet of poultry. However, use of unstabilised bran in poultry feed is risky, for unsaturated acids in the process of becoming rancid may lead to loss of tocopherols and hence to vitamin E deficiency. Up to 60% bran can be used in the ration of dairy cows.

9.6.4   Use of rice bran for human food

Considering the composition of rice bran and the contents of valuable nutrients in it, rice bran would appear to have a natural potential to be used as human food. Indeed use of rice bran as a possible food ingredient has been long thought of and researched upon. Dr. Salvador Barber of Spain in a review article in 1974 wrote "everything would indicate that the possibility of using rice bran as food is imminent". However, this possibility has yet to be fully realised four decades later.

The 'polish' obtained during further 'polishing' of milled rice as in the Western countries contains less fibre and ash and more starch than bran but essentially similar protein as bran. It has been used to some extent in foods, especially in baking and in some baby foods. However, bran as such with or without purification has not yet been widely used.

Use of bran in baking for bread, cakes, pies, cookies and biscuits as well as in snacks and confections has been suggested, but such use has been limited. Bran can substitute up to 10% of flour for baking of bread; anything more affects loaf volume. Protein concentrates obtained from bran as described below have the potential of being used in various products. Milk-like or beverage-like products have been made from bran either by water extraction of full-fat bran or from the concentrates mentioned. The functional properties of rice bran proteins are of major importance in possible use of bran proteins in food formulations. These properties (solubility, water absorption, emulsification, fat absorption, viscogenicity, foaming) have been found to be generally satisfactory.

The major impediments to the use of bran as human food are:

   i. the instability of the bran oil because of the lipase enzyme,

   ii. the high amount of fibre in bran, especially insoluble fibre,

   iii. its excessive phytin content,

   iv. its unbalanced C:Pa ratio, and

   v. the often unclean condition of the bran (too high sand and silica).

Among these, the first difficulty (instability of bran) has been solved by the modern process of stabilisation. The third and fourth problems are problems of improper nutrient balance that can be largely taken care of by adjusting the would-be food composition. The last difficulty (unclean bran) is ultimately a question of using proper technology, including that for cleaning of the paddy during milling.

But the problem of excessive fibre, especially insoluble fibre, is a big hurdle. Attempts to solve this problem have taken the following lines:

   (a) dry fractionation of the bran to eliminate or reduce the insoluble fibre content,

   (b) solubilising the bran by enzyme treatment, and

   (c) wet fractionation of the bran to obtain enriched fractions containing most of the protein and other valuable nutrients.

Several attempts have been made to obtain enriched edible fractions by physical dry fractionation of the bran. Sieving, grinding and air classification in different combinations and in different degrees have been tried to obtain a fraction which is enriched in protein and other nutrients but having a much lower fibre content. These attempts have not been very successful. Even if a reasonably balanced and nutrient-rich fraction has been obtained, the proportions of the original nutrients thus recovered have not been satisfactory. Attempts have been made to solubilise a greater proportion of the bran by enzyme action. Various carbohydrases (including cellulase, hemicellulase, amylase, etc.), proteases and phytase have been used. The resulting mix is then dried, giving a highly nutritious edible powder. This approach seems to have a good potential. But how far it will be cost effective in practice remains to be seen.

Some success has been achieved with the final option, viz. wet fractionation. Both alkaline extraction and plain water extraction have been tried. Solubility of bran proteins increases with increasing pH (that is, alkaline). This is the basis of two or three systems in which rice bran is extracted with dilute alkali at a pH ranging from 9 to 11.5. Solubility increases with increasing pH. But pH of over 11 is hazardous for it reacts with lysine and may produce a potentially toxic compound with serine. The solution is separated by centrifugation.

The protein is then precipitated with acid at the isoelectric pH of 4–5. Some more protein can be recovered from the solution by coagulating with heat. The material is dried in a drum dryer. Protein concentrates having up to 85% protein can be obtained in this way, recovering up to 50% of the bran protein. The concentrate has essentially the same amino acid composition as in the original bran and good PER value.

In an alternative approach, rice bran suitably ground in dry or wet state is extracted with water. The slurry is either decanted or screened to remove the fibre. Suitable treatments of heating, centrifugation and screening can give three fractions : (i) a high-fibre (18%) intermediate-protein (13%) residue, (ii) a low-fibre (< 3%) higher-protein (16–20%) fraction, and (iii) a syrup containing the water-soluble vitamins and other nutrients. The second fraction can be used in food and the first in feed.

Recent developments in the industry

Despite many research studies along various lines over the decades to use rice bran as human foods, as briefly outlined above, their practical application by the industry have so far been not encouraging. However, there are signs that this scenario may change. Utilisation of bran for food seems to be attracting the attention of food industry now. Recently at least one company in the USA and one or two in Japan have taken considerable steps to prepare various food ingredients from rice bran. These attempts have taken various lines such as the following:

  i. Rice bran, once stabilised, is a rich source of protein, dietary probiotic fibre, nutritious oil, carbohydrates, lecithin and various micronutrients (vitamins, minerals, antioxidants, oryzanol, ferulic acid, phytosterols). It is for the manufacturer to keep this factor in mind and so process the bran for one product as to try to simultaneously recover other valuable products as coproducts or byproducts.

  ii. Rice-bran oil has many highly attractive properties. For one thing, it has a high smoke point and is therefore very suitable for use as a frying oil. Its nutritious properties are well known. It contains gamma-oryzanol, tocopherols, sterols, antioxidants, ferulic acids, etc., whose physiological effects are well known. The oil is thus hypocholesterolenic and hypoallergenic, it protects from ultraviolet rays and skin damage, is anti-ageing and is helpful in fighting or alleviating Alzheimer's diseases, tumour and high blood glucose.

  iii. Appropriately treated rice bran improves stabilising and texturising properties of meat emulsions and coarse-ground meat systems such as hamburgers and chicken patties. It can act as a 'functional filler' in processed meat products, enhancing and stabilising meat emulsions and coarse-ground meat. Rice bran has significantly good functional properties, including good solubility in water and fat and good viscosity. It enables retention of moisture, so tends to enhance the shelf life of products. It can help to replace carrageenan, modified starch, etc. because of its viscogenic properties. It can also help in modulating viscosity. It can be a good replacement for soy and other vegetable proteins because of its easy hydrating and diffusing properties without lumping.

iv. Wet fractionation system has enabled production of a 35% protein concentrate, a 90% protein isolate and a protein hydrolysate. Apart from other use, their emulsion and foam-stabilisation properties can be useful in producing rice milk, prepared food, baked food and extruded meat analogues. This can help in sports performance, weight management, blood glucose management, and that of cholesterol.

Further reading

The science and technology of rice bran and its utilisation have been discussed in detail in all text books/chapters of books on the rice grain, mainly the following:

- Luh (1980)
- Juliano (1985)
- Champagne (2004)

In addition, the matter has been extensively reviewed in:

- Barber and Benedito de Barber (1985)
- Saunders (1990)
- Fabian and Ju (2011)
- Sharif et al. (2014)
- Friedman (2013).

Recent developments in the industry can be found in the following book and the two web sites below :

- Hoogenkamp (2012)
- http://www.tsuno.co.jp
- http://www.ricebrantech.com

Rice husk and its utilisation

Rice husk, also called paddy husk or hull, is the major byproduct of rice milling process and the industry. It constitutes, on an average, about 20% by weight of paddy. With world paddy production around 745 million tonnes (FAO 2014), and around 80% of it being converted to milled rice, there is an estimated generation of about 120 million tonnes of husk the world over. This is the highest quantity of a byproduct from any industry in the food sector. Utilisation, or even disposal, of this low-value byproduct has been a challenging problem. Use or disposal has frequently proved difficult because of the tough, woody, abrasive nature of the husk, its low nutritive value, resistance to weathering, great bulk and high ash content. The problem is magnified due to its frequent accumulation and piling up at the mill site. As against the other byproducts emanating from the mill, which keep getting cleared from the mill premises continuously, husk generated generally keeps accumulating. If there is no ready off take, it poses problem of disposal. By far its traditional major use has been as boiler fuel for the production of steam in rice mills and parboiling plants. Not that attention on exploring ways and means to utilise this byproduct have been lacking. Studies on its structure, composition, properties and possibilities of its use in various industries have been reported over many decades. A good amount of literature has been generated through a large number of publications on varied potentialities for its uses, mostly at laboratory level. But commercial usage has been rather limited due to many factors. The situation, however, seems to be changing now, as many new applications are being tapped for its economic and commercial exploitation.

10.1   Composition and properties of rice husk

The proximate composition of husk is presented in Table 10.1. It could be seen that husk has a very low content of protein and available carbohydrates, but a high content of crude fibre, lignin, and crude ash with silica in it. These components, particularly the high content of lignin and ash, make rice husk the cereal byproduct with the lowest digestible nutrients, lowering its value even for feed purposes. However, the high content of ash (and silica in it), and of pentosans render it suitable for commercial exploitation, as discussed later.

Table 10.1 Composition of rice husk (% by weight) [*Source:* Juliano, 1985; Reproduced with permission from AACC International]

Protein	1.9–3.0
Fat	0.3–0.8
Crude fibre	34.5–45.9
Available carbohydrates	26.5–29.8
Lignin	9–20
Ash	13.2–21.0
Total digestible nutrients	9.3–9.5
Cellulose	28–36
Pentosans	21–22

Some important physical properties of husk, which are relevant to its handling, conveying, storage and utilisation, are shown in Table 10.2.

Table 10.2 Physical properties of rice husk [Compiled from Houston, 1972]

Equilibrium moisture, %, at 50–60% RH	9.1–10.5
Bulk density, gm/ml	0.1–0.16
Bulk density when compressed, gm/ml	0.4
True density, gm/ml	0.67–0.74
Thermal conductivity, W/(m.°C)	0.0359
Fuel value, kcal/kg	3300–3600
Angle of repose	35°
Angle of repose of ground husk	43–45°

It may be noted that the husk has a very low bulk density. For equal weight it would occupy about five times the volume as that of paddy, and about seven to eight times as that of rice. This makes it a costly proposition for its transport. Compressing it increases the bulk density up to three times, while grinding, by about four times to reduce its bulk. This helps to improve the situation. It is thus advisable to use it, as far as possible, at the mill site or any other place nearby. Its thermal conductivity is very low, which is comparable with thermal conductivity of the excellent insulating materials, showing its potential for exploitation as a thermal insulator based on this property. Its fuel value, for which the husk has largely been used so far, is nearly two-thirds of bituminous coal and about one-third of oil/petrol. However, the burning system has to be very efficient to tap this calorific value completely. Normally, some quantity of unburnt carbon remains in the residue by using systems which have been traditionally employed for burning.

10.1.1   Utilisation of husk

In most of the major rice producing countries, the total annual output of husk is very large, but most of the husk is available in relatively small quantities at tens of thousands of small mills spread in these countries. Depending on the size, type and location of the mill, the situation with regard to utilisation or disposal of the husk is different from country to country, and location to location in the same country. The problems faced by large modern mills located in relatively urban areas are very different from those of small mills scattered in numerous villages. However, numerous studies have been reported in the past few decades, exploring various ways and means towards utilisation of this voluminous byproduct from rice milling industry. A list of uses as given in one of the reviews on the subject, published in 1971 is given in Table 10.3. The list has enlarged presumably further with many more publications added since then. While some of these are being commercially used, some others are shown as potential only at the laboratory level. Still others would form an insignificant part in the total usage scenario.

Table 10.3 Areas of utilisation of rice husk [Compiled from Houston, 1972; Reproduced with permission from AACC International]

1.	Abrasives	14.	Glass manufacture
2.	Absorbents	15.	Hydroponic media
3.	Additive in manufacture	16.	Insulation
4.	Bedding	17.	Litter and nesting material
5.	Building material component	18.	Packing material
6.	Carbon source	19.	Panel board
7.	Carrier	20.	Pigment
8.	Cellulose pulps	21.	Pressing aid
9.	Feeds	22.	Raw material source
10.	Fertilisers	23.	Refractories
11.	Filter media	24.	Soap manufacture
12.	Fuels	25.	Silica source
13.	Filler material	26.	Soil amendment

10.2   Agricultural uses of rice husk

10.2.1   Feed

Due to its very low total digestible nutrient fraction (<10%), but a high content of silica and lignin, rice husk has a very low nutritional value as compared to other feed materials. However, husk is incorporated in cattle feed after grinding, up to 20% level, on functional basis to provide roughage in the cattle feed. Husk ground to different mesh size is marketed for this purpose in different countries. Usually, husk ground to pass through 20 or 30 mesh sieve, but retained on 80 mesh sieve, is used for compatibility with other ingredients. Fine husk, passing through 80 mesh sieve is also sold for incorporation in feeds that are palletised. In this case it acts as a binder also. Incorporation of ground husk from 25% to 80% level in the sheep feed did not cause any detrimental effect as reported from studies in India. In USA, a 'mill feed', consisting of all byproducts from the rice milling, combined in the proportion produced, is also sold. This consists of 61% ground husk, 35% bran and 4% polish. The mix is sold under guaranteed analysis such as more than 5% fat and 6% protein, and less than 16% ash and 30% crude fibre. This product is similar to the 'huller bran' that is produced from huller mills prevalent in rural areas of developing countries. It is a common practice to use the 'huller bran' as cattle feed in these countries. The 'huller bran' normally consists of 60% husk, 30% bran and 10% polish, germ and broken rice. Rice mill byproduct is commonly sold as poultry feed in Asia.

Treatment with alkalies has been used either to remove the silica (with potential silicate production), or to modify silica and other components, in order to improve digestibility. Use of steam under pressure, in addition, was shown to increase the digestibility further. The USA Food and Drug Agency has approved use of such feed supplement at not over 20%.

Pretreated rice husk has been used as a substrate for microorganisms producing single cell protein. A fermented rice husk feed for livestock and poultry has been developed in Japan after grinding and subjecting it to heat and alkali treatment which increased the protein content to more than 16% in the feed.

10.2.2   Bedding and litter

A small quantity of husk is used traditionally as bedding and litter material for cattle, horse, sheep, poultry and laboratory animals. In Philippines and Indonesia husk is used in duck hatcheries mainly for insulation purpose. Many advantages have been listed for using rice husk for animal bedding in comparison to other materials. It is fire resistant. It does not attract insects. It is light weight and spreads over a larger area than saw dust or other materials for the same weight. It does not absorb liquid and bind into clumps. Urine and feces percolate through the husk to the lower level, keeping the animal dry and making cleaning easier and faster. The soft bedding also prevents hock and leg blisters in animals.

10.2.3   Soil treatment

Although rice husk does not have much of fertiliser value, its incorporation in soil has been shown to make the soil phosphorus and silica more available to the rice plant. This increased the resistance of the plant to lodging and the insect attack. Even higher yields of rice, up to 6%, have been reported by use of husk in soil preparation at the level of 10 to 30 tonnes per hectare, although the growth of the plant was slightly retarded. Another large use of husk, and husk ash, is as diluent and anti-caking agent in commercial mixed fertilisers. Rice husk and husk ash also find considerable use in mulches in nurseries and gardening. In some Asian countries, like in Japan, husk is also used to form 'husk pipes' as a mean for channelling water underground in fields. The porosity of husk allows water to flow through, while the physical bulk remains undisturbed.

10.3   Industrial and other direct uses of rice husk

10.3.1   Fuel and energy

The largest use of the rice husk has been as a fuel, as a source of energy in many areas of the world where rice is produced. In the industry, it has been used for the production of several types of energy as required. This includes industrial process heat in the rice milling and other industries, mechanical and electrical power for rice milling, and even electrical power for the grid. The burning systems used are, direct combustion in rice mill furnaces, in boilers for the production of steam and to run steam engines, and gasification to obtain producer gas to run turbines producing electrical power.

Industrial process heat

At present the most common industrial use of rice husk is for process heat in the rice milling industry, including both the direct use of heat from husk-fired furnaces and the generation of low pressure steam in husk-fired boilers.

As a fuel in furnaces, it is used especially in mills producing parboiled rice in the Indian sub-continent. In India, for example, as nearly 50% of rice is parboiled, the magnitude of husk utilised in these mills could therefore be emphasized. Husk is used for direct combustion in furnaces for the generation of steam in boilers, for heating water, for steaming of soaked paddy, and for hot-air drying of parboiled paddy. Hot air for drying purposes is generated by direct or indirect means. In the direct method, air is mixed with flue gases, whereas in the indirect method, the gases are passed through heat exchangers to generate hot air required for drying. Direct method has higher efficiency, but indirect method results in cleaner hot air. As high pressure steam is not required for processing in these plants, low-pressure boilers are normally employed, that produce less than 10 bars pressure steam. Combustion of husk in parboiling and milling plants has the advantage that husk is consumed where it is produced and only the ash requires disposal. On account of this, India is the largest user of rice husk, burning about 10 million tonnes of husk annually in the rice milling industry alone, as more than half of rice produced is parboiled in the country.

The efficiency of burning the husk and the realisation of its heat value depends on the type of furnace and the technology involved. In majority of the installations in the south Asian countries, furnaces used are of inclined grate type, in which the burning efficiency is rather low (about 40–50%) as some amount of carbon remains unburnt. Partial and uneven combustion also leads to smoke and some fly-ash emission, causing environment pollution. Char produced from traditional rice-husk furnaces may contain about 35–40% organic matter, mainly carbon, due to incomplete combustion. The high silica content of husk, and the way it is distributed in it, gives it a rigid, skeleton-like structure. Because there is non-uniformity that occurs in the fuel-air ratio in the grate-type of furnaces, it leads to significant amount of unburnt carbon in the ash. Coupled with this, there is a considerable amount of carbon that remains trapped in the rigid ash-skeleton and cannot be burnt. However, use of high efficiency furnaces like cyclone furnace or the fluidized-bed combustion furnace, is now on the increase, in which the husk is kept agitated under suspension while burning. The turbulence due to fluidisation in the bed breaks the rigid ash-skeleton to make the trapped carbon available for oxidation, thus ensuring its complete burning. Further, other improved operational parameters in these furnaces make them a better choice for burning of husk.

Apart from rice mills, husk is also used as fuel for boilers for steam generation in rural-based industries in Asian countries. Generally, use of Lancashire type of boiler has been more popular in such cases, although there is a trend to change to the more efficient fluidized-bed combustion boilers. The steam generated is used for running steam engines for mechanical and other power requirements of the industry concerned.

Gasification

Manufacture of producer gas by burning rice husk deserves a special mention, for it is the system which is being adopted very widely for utilisation of rice husk. Gasification is a process in which solid biomass is converted to gaseous fuel by heating it in a limited supply of air, oxygen or steam. Unlike combustion, where oxidation is substantially complete in one process, gasification first converts the intrinsic chemical energy of the carbon in the solid fuel into a combustible gaseous phase containing methane, carbon monoxide, hydrogen as well as small quantities of inert gases. Certain amount of char and tar is produced in the process, which is removed. The clean producer gas is then used directly for running a gas turbine or for producing high-pressure steam to run a steam turbine for generation of electrical power. The exhaust steam is utilised for the process-heat requirements for the industry. The producer gas from the gasifier has a calorific value of 1450 kcal per cubic meter. About 1.5 to 2.8 kg husk gives one kWh of electricity.

Gasification is considered as a clean technology as there is almost no emission of carbon dioxide and other gases that contribute to global warming. There is a complete burning of the gases produced. Further, the fuel is the agricultural waste, thus making use of the renewable resources. Simultaneously, utilisation of the waste biomass takes care of the disposal problem. It is therefore considered as an eco-friendly, clean and green technology. On account of this gasification is encouraged by governments in many countries with incentives.

There are several types of gasifiers with different designs of the unit (also called reactor or the combustion chamber) that are used. Examples are: fixed-bed gasifiers with updraft or down draft, with straight tubular design or with a neck in the middle, fluidised-bed gasifiers or the bubbling-bed gasifiers. Each design has its pros and cons. However, the down draft system is more popular as it is comparatively cheap and produces very low tar content. It suits very well also for the small-scale bio-mass plants based on husk. In the down-draft gasifier unit, the husk is fed at the top through a hopper. As it moves down slowly, it is set alight. Air is drawn downwards through the rice husk, which burns it in a restricted supply of oxygen. Under the heat in the reactor, the chemical energy of the husk is converted to a combustible energy-rich producer gas in a complex series of oxidation, reduction and pyrolysis reactions. The char and ash that remain, get dropped to the bottom of the gasifier and are removed.

A particular point in favour of rice husk being highly suitable material for gasification is worth mentioning. All other agricultural waste materials, which normally come in varying sizes and shapes, need to be processed for bringing them to a more or less uniform smaller size before feeding in to a gasifier. Rice husk, on the other hand, actually comes in such a tiny and uniform size and shape that it is ideally suited for burning to produce the producer gas. Each husk particle is boat-shaped. The boat shape and uniformity in the husk particles entraps certain amount of air when the husk occupies a volume. The air is thus automatically limited but spread very uniformly inside the biomass volume. This promotes a slow burning, or smouldering, of the biomass and releasing the gases which finally burn giving a smoke-less blue flame.

Mechanical and electrical power

Apart from production of steam for process heat requirements in rice mills, husk-fired furnaces and systems have been developed in various countries, like USA, Italy, Germany, Belgium, Japan, Brazil, Thailand, China and India, to generate mechanical power and electrical power as well. These systems generate the power through combustion and/or gasification, either as cogeneration or stand-alone units. Such systems are working in many of the rice producing countries. If the cogeneration system is just enough to meet the requirements of the rice mill, it is a "zero energy" mill. Such plants for small rice mills are offered by manufacturers in China, Thailand and India. For example, Grain Processing Industries in India offer their initial model that matches the needs of a 2 tonnes per hour rice mill, processing parboiled rice. The entire husk (about 400 to 420 kg per hour) of the mill is used in the gasifier. If the mill is not engaged in parboiling, the producer gas is used to run a generator to produce 220 kW power. As a 2 tonne per hour mill would need only about 60 kW, the extra power could be sold to another industry or the grid, bringing in revenue to the mill. Such systems to suit to 3 tonne per hour rice mills have also been developed and supplied across Thailand and Philippines by a Thai company and in Malawi by a German firm.

At the other end of the spectrum, generation of electric power by burning husk has been successfully practised in very large mills, where husk is generated in large quantities at a single spot, or available within a radius of a short distance. The power is not only used for captive requirements, but is also sold to the grid or other industries. Producers Rice Mill Energy Systems, Arkansas, USA, have installed two units during 1982 at the Producer Rice Mill, Arkansas, which have been used for providing process steam in parboiling and drying of rice and also for power generation. These units burn 2 to 2.3 tonnes of husk per hour, with gas temperatures around 1000°C in the reactor. Gases from one of the units are used, after appropriate mixing with ambient air, for drying of parboiled rice directly, and the gases from the other unit are used to generate 160 kW of electric power for use in the mill. The company has also installed a few more 225 kW power plants in rice mills in Malaysia during 1990s, to match the needs of 5 tonne per hour rice mills.

Another large rice mill viz., Farmers Rice Milling Co., in Louisiana, USA, has a husk-fired cogeneration plant, supplied and installed by Agrilectric Power Partners, Arkansas, that has been producing 12.5 MW electric power since 1984. The Mill processes about 1500 tonnes paddy per day and the entire husk produced (about 300 tonnes per day) is ground and burnt by the cyclonic suspension burners in the boiler furnace. This produces superheated, high-pressure steam at 44 bar and 400°C that generates electric power through steam turbine system. The exhaust steam is used for process heat requirements. A part of the power produced is used for the mill operation and another facility near the mill, but a larger part is sold to the power grid.

Another large husk-fired power plant, perhaps the world's largest husk-to-energy plant, is located at the outskirts of the city Williams in California, USA. The facility was created in 1990 to solve the disposal problem of husk produced by many rice mills in the Colusa County of California, which contributes to about 20% of the US rice crop. About 700 tonnes of husk is burnt per day (about 29 tonnes per hour) after grinding, by direct combustion, in an improved furnace at about 860°C. The high pressure superheated steam runs high-efficiency turbines that drive a 30 MW generator. The net production is 26.5 MW electrical power. It is a stand-alone commercial venture, not attached to any rice mill. Many large rice mills that exist in a relatively small area provide the husk for the plant. The power generated is sold to an electricity supplying company of the area.

The largest power plant in any developing country using rice husk as fuel is located in the State of Punjab in India, attached to perhaps the largest rice mill in the world. The rice mill, a facility of KRBL Ltd., processes 3600 tonnes of paddy per day, generating more than 700 tonnes of husk per day. The husk is burnt in a system, manufactured and supplied by a German firm, by direct combustion in an improved grate furnace. The high pressure steam of 72 bar, generated through boiler, and super-heated to 505°C, runs a two-stage steam turbine producing 12.3 MW electric power. The exhaust steam meets the process heat requirements of the rice mill. Twelve such husk-fired cogeneration plants, with capacities of 10 to 16 MW electricity output, manufactured by different firms, have recently been installed in rice mills in the rice belts of Haryana and Uttar Pradesh in India. One of them, installed in a rice mill complex during 2010 at Mathura, Uttar Pradesh, processing 90 tonnes paddy per hour, produces 16 MW electrical power.

A Malaysian and Vietnamese joint venture company is building a rice-husk based power plant of 2 MW capacity, at a rice mill in Hua Giang, Vietnam, during 2014. This would be the nation's first husk-based power plant and would consume 250 tonnes rice husk per day. Twenty more such plants are planned in the next 5 years. Lack of government incentives has been cited as the reason for shunning so far the power plants based on renewable energy.

Satake Engineering Co., Japan, has supplied many husk-fired power plants in Myanmar, Malaysia, Thailand, and the Philippines, designed for rice mills processing from 7 to 20 tonnes of paddy per hour. The net output of these plants ranged from 288 to 1200 kW.

In general, a mill that uses most, or all, of its husk output in a low-pressure power plant produces power which is needed to operate the mill, leaving little or no surplus power for export. Net power output of 275 to 350 kWh per tonne of husk is equivalent to 55 to 70 kWh per tonne of paddy. This level of output may be close to, and can be less than, the power consumption per tonne of paddy of a large modern mill, in which, power is required for many additional machinery in the processing line. The total power requirements of very large rice mills often include increased power requirement for lighting, housing water pumping, office and other infrastructure.

Increasing cost of the fossil fuel and shortage of electrical power, particularly in the developing countries, has led to an increase in the demand of biomass-based small-scale gasification power plants. Rice husk therefore is not a waste any more. More than 100 suppliers from different countries, of which a large number is from China, now offer a wide range of husk-fired small cogeneration gasifier power plants with rated outputs from 50 kW to 6 MW.

In an interesting recent development in India, Husk Power Systems, a company based in Patna, in the state of Bihar, has started manufacturing and installation of mini husk-fired gasifier plants in rural areas. The producer gas from the reactor, after cleaning, is burnt in an internal combustion engine to generate 25 to 100 kW electrical power. These units are meant for rural electrification, where there is no supply of electricity through power grid. Electricity is supplied on a pay-for-use basis through these plants to villages and hamlets of up to 4000 inhabitants, within a radius of 1.5 to 3 kilometres, by establishing village-based micro grids. More than 90 such mini power plants were installed till February 2013, in the states of Bihar and Uttar Pradesh, changing life in the remote villages of India. Rice husk for these plants is collected from mills located within a radius of about 10 kilometres.

Burning of husk for fuel purposes, in all cases, leaves ash as the residue. This is an important and useful byproduct. The subject of utilisation of rice husk ash is discussed later in a separate section.

Briquettes

Briquettes are a compressed form of biomass, in convenient size and shape, produced from agricultural or agriculture-based industrial waste. These are used as fuel for domestic as well as for industrial purposes. They provide a cheaper substitute for coal and charcoal. Size reduction and compaction of biomass increases their density many folds (in case of husk up to 10 times), which makes their handling, storage, transportation, and marketing much easier and economically viable. Many manufacturers from Vietnam, Thailand, India, Malaysia and China offer briquetting machinery, as well as rice-husk briquettes.

Combining partially bunt husk along with fresh husk for briquette making, to increase calorific value of the briquettes up to about 3800 to 4000 Kcal per kg, is also practised. Rice-husk briquettes made of powdered husk charcoal are also marketed, sold as barbeque or hookah charcoal. These have carbon content of up to 90%, with calorific value of about 8500 kcal per kg and provide smokeless burning.

Domestic stoves and kilns

Special stoves, which use rice husk as fuel, have also been designed and are marketed for use in kitchens. These are common in many rice-producing developing countries like Thailand, Indonesia and Philippines. These stoves provide smokeless cooking with blue flame, thus eliminating eye and lung diseases due to smoke by burning wood, which is common in rural areas. They provide better hygiene and could be fabricated at rural level.

Another traditional use of rice husk is its use as fuel in brick kilns in the developing countries. Burning of husk in kilns ensures supply of a slow but continuous high-temperature heat over a long period, which is required for producing good quality bricks.

10.3.2   Building and construction materials

Husk-based panel boards

Wood and wood-based panels and particle boards are used for various purposes in the construction industry. However, with the growing shortage of wood and its increasing cost, demand for light-weight and high-strength composite boards from alternate sources is also increasing. Rice husk has become a substitute material for replacement of wood and wood-based board products.

Various types of board panels are produced using rice husk, such as particle boards, insulation boards and ceiling boards, for different requirements of the intended applications. Husk-based particle boards are produced by spraying rice husk with a thermosetting resin, or an appropriate adhesive, in a rotating mixer, followed by compression in a hot hydraulic press. Densities up to 800 kg per cubic metre could be achieved. Low-density boards, having densities between 300 and 400 kg per cubic metre, are used for thermal and acoustic insulation purpose, whereas higher-density boards are used for other purposes.

Husk is ground to fine powder in order to make better quality boards. For getting appropriate properties in the boards, partial replacement with wood fibre, chips, saw dust and sugar cane bagasse is also practised. In fact, such composites have better properties, than husk alone, in term of utility and replacement of wood-based particle boards. Alkali treatment and saw dust inclusion increases the tensile strength of the board. Such boards are used for ceiling and roofing work.

Different types of adhesives and resins are used as binding agent in the manufacture of the boards. Commonly used synthetic adhesives are phenol-formaldehyde and urea-formaldehyde. The former creates strong and water resistant bond, but is costly, while the latter is cheaper and gives better surface smoothness, but has lower water resistance. Low-cost natural adhesives, like starches, modified starches and alkali-modified soybean concentrates are also used where exposure to water or moist atmosphere is not intended. The boards can be painted for decorative finishes. They can also be glued to each other or to decorative laminates using suitable adhesives. These boards can be used for architectural purposes in building interiors such as wall or ceiling linings, panelling, partitioning and for providing insulation in buildings. Proneness to ignition is also low for rice husk particle boards.

A few manufactures from India, Vietnam and Thailand have included such boards in their supply, in addition to the normal wood-based particle boards.

Insulating bricks

Rice husk has been used in the manufacture of porous and light-weight insulating bricks. For this, rice husk is mixed with clay in making the bricks. When these bricks are burnt in the kiln, the husk burns out, resulting in a porous structure of the bricks. Presence of entrapped air in the pores renders the brick to have thermal insulating characteristics.

10.2.4   Miscellaneous uses

Filter aid

A small quantity of rice husk is utilised as a pressing aid for noncitrus fruits. The husk is added at 1% level by weight of the fruit, and the mixture is pressed. The husk actually serves as a channelling or flow aid, not as a filter. This is done for a variety of fruits like apples, prunes, berries, cherries, pears, and grapes. It is claimed that this gives greater juice yields, drier pomace, and reduction in pressing time. However, use of husk for this purpose requires a prior thorough cleaning. Husk from the regular mills, processing raw paddy, is washed and sterilised. Alternatively, husk from parboiled rice is also used. Washing or parboiling reduces the chance of introducing unwanted microorganisms.

Abrasives

The rice husk has a rough and hard surface. This property is made use of for making abrasives. Husk is used either in the form of whole husk or after grinding. Whole husk may be used for tumble-cleaning and polishing. The action produces brightly polished surfaces in barrel-tumbling of iron, aluminium, brass or bronze parts, better than that produced by use of sand. This type of use is also effective in polishing of small plastic parts and gem stones.

Ground husk, mixed at 40% level with corn cob grits (another abrasive agent), has been shown to give better performance for soft-grit blast-cleaning of machined engine parts than the cob grits or husk alone. The mixture is used for cleaning carbon, dirt, and scales form aeroplane cylinders, pistons and other parts rapidly, cleanly, and economically without removal of metal or disturbance of tolerance in closely machined parts. This use seems to be expanding.

Rice husk, ground to pass through 20-mesh screen, but not 100-mesh, has been used as an abrasive material which is added to mechanic's hand-soaps. Ground husk, finer than 100-mesh, no longer feels gritty. The mechanic's hand-soaps normally contain about 10 to 25% abrasive material.

Chemicals

Rice husk has been reported as a raw material for producing a number of organic compounds in the chemical industry. Examples are: furfural, lignins, nitrolignins, oxalic acid, levulinic acid, coumaric acid, saccharic acid, acetic acid, sulphonic acid, etc. Hydrolysate of rice husk has been used for growing yeast or as substrate for various fermentations. Most of such specific uses, however, have products of limited market, where economics, logistics and other considerations become limiting factors, not the technology.

Elemental silicon for electronics

Recent researches from Stanford University, California, USA, and also from a Korean University, have opened a new potential for a high-end use of rice husk, promising to be a big boon for the electronic industry. A bench-scale process has been developed to directly extract elemental silicon-nanoparticles from rice husk. These are 10–40 nanometres in size and have a highly porous structure. Lithium-ion battery anodes made from this material exhibit high performance, with very high reversible capacity, seven times greater than the presently used graphite anodes. Further, these anodes have a long cycle life and show 86% capacity retention over 300 cycles, as against decay within 100 cycles in the present batteries.

Present methods for the production of nano-silicon anodes involve high-temperature (about 2000°C) pyrolysis of its precursors (like high purity quartzite) or laser blasting of bulk silicon. Both are high-tech, energy-intensive, costly processes. Using rice husk as the raw material source, it is thus possible to have a large-scale, overall energy-efficient, low-cost synthesis of highly functional silicon nanomaterial from a renewable source. If commercialised, it will have a great impact for the portable electronics and electric car industry in future.

Other applications

A firm from California, USA, and a multinational firm based in Australia have started manufacturing and marketing biodegradable tableware products made from rice husk under the promotion of 'green technologies'. These include plates, bowls, glasses, cups, forks, and spoons etc. Another company from Spain is manufacturing and marketing chopsticks made from rice husk. In China alone, an estimated 45 billion pairs of wooden disposable chopsticks are used and thrown away every year, the equivalent of almost 4 million fully grown trees. Japan is another major consumer of disposable chopsticks. To preserve its forests, Japan imports disposable chopsticks from China, Vietnam, Indonesia, Chile, and Russia, resulting in deforestation in those countries. Japan, however, still has to deal with the problem of used chopsticks—an estimated 25 billion pairs every year. This is where rice husk comes in.

Husk is also used as pillow material in some parts of the world. The pillows are loosely filled and are considered as therapeutic. Some firms are producing and marketing rice-husk pillows in Thailand and China. The husk-filled pillows maintain the desired shape and curve and are advised for therapeutic use. These pillows are becoming very popular in China.

10.4   Rice husk ash (RHA) and its utilisation

Burning of husk as fuel results in the production of ash as the residue or the byproduct. Rice husk ash (RHA) generally consists of about 5–30% carbon, and about 70–95% silica, and less than 1% alkali oxides and alkali earth oxides. On account of this, utilisation of RHA has been focussed to tap its carbon potential, when it is in the form of char (ash colour is black) or its high silica content when the carbon content is very low (ash colour is grey, pink or white), as discussed below. The variation in the carbon content is caused by the amount of air used during burning. Low amounts of air in contact with the husk during combustion produce ash with higher carbon content. On the other hand, if an ash with high silica content is desired, large volumes of air in close contact with the husk particles must be maintained during combustion. Further, the temperature of combustion has a major effect on the nature of silica that is present in the resultant ash as discussed later.

10.4.1   Char and activated carbon

High-carbon RHA (char) has been shown to have a variety of uses. As mentioned earlier, it is ground and compressed, with or without raw husk to produce briquettes, barbeque coal or similar products with high fuel energy value.

Char, in conjunction with shredded coconut fibre has also been used as a simple, cheap and efficient way to treat water for rural communities, as demonstrated in Thailand. For this, surface water from pond or stream was first passed through a bed of shredded coconut fibre followed by that of rice husk char. The treatment reduced turbidity, removed bacteria, and produced water with acceptable taste, colour, and odour.

For better performance of carbon, char - or any other high-carbon material, like coal, wood, coconut shell etc., which are the traditional sources of carbon - is activated by physical or chemical treatments. Physical activation is performed by heating the material at 750–900°C in air, steam or CO_2, while chemical activation is carried out by treatment with a strong acid, a strong base, or chlorides of zinc and calcium, at relatively lower temperatures of 400–600°C. Chemical treatments are generally preferred as they result in creating micro pores in the material with higher surface areas, even up to more than 2000 square metres per gram. Many of the uses requiring absorption and adsorption phenomena depend on the surface area of the material, and its anionic or cationic activity. Char from rice husk has a macroporous structure with rudimentary surface area (less than 50 square meters per gram), which limits its use for such purposes. Conversion to activated carbon increases its surface area manifolds (500 to more than 1000 square metres per gram), thus imparting it a high adsorption capacity. Husk-based activated carbon is a new entrant in this field and has to compete with the standard commercial ones.

Activated carbons are used for a wide range of industrial and residential applications, including municipal and industrial wastewater treatment, residential tap-water purification, industrial air purification, food and pharmaceutical purification and solvent recovery.

One advantage, which makes the husk-based activated material special, is that it contains two adsorptive components, a carbon-rich char and a high content of reactive silica in it. The porous carbon derived from rice husk has also fast kinetics and appreciable adsorption capacities for use in new applications like catalytic support, battery electrodes, capacitors and gas storage. As the market of activated carbon is very large, it is expected that husk-based activated carbons would stay in the field.

10.4.2   Low-carbon and carbon-free ash

Low-carbon (grey) or carbon-free (pink or white) ash can substitute equally well for high-carbon (black) ash in many potential applications, except where black colour or its heat-absorbing quality is desired. Composition of carbon-free ash is presented in Table 10.4. In fact the low-carbon or the carbon-free RHA has much more applications and a wide variety of uses than the high-carbon RHA. Essentially, all these application depend on the silica and the state in which it is present. As mentioned earlier, burning of husk at lower temperatures, at about 400–700°C, yields ash in which amorphous nature of silica, with high porosity and reactivity, is preserved. Burning of husk at higher temperatures, on the other hand, leads to structural transformation of silica by formation of crystalline silica, at around 900°C, and to sintering (glass formation) beyond it. This leads to the loss of porosity, its surface area and consequently functionality for some high-tech applications.

Table 10.4 Composition of rice husk ash, % dry basis [Compiled from Houston, 1972; Reproduced with permission from AACC International]

Constituent	Value (Range)
SiO_2	86.9 – 97.3
K_2O	0.58 – 2.5
Na_2O	0.0 – 1.75
CaO	0.2 – 1.5
MgO	0.12 – 1.96
Fe_2O_3	trace – 0.54
P_2O_5	0.2 – 2.85
SO_3	0.10 – 1.13
Cl	trace – 0.42

Rice husk ash has been utilised as a source of amorphous silica which finds wide applications in various chemical industries, such as ceramics, glass making, steel production, pharmaceuticals, rubber, plastics, refractory materials, cement, paints, soaps and detergents, polymer composites, electronics, refining of vegetable oils etc. Some of these uses are discussed below.

Steel industry

Rice husk ash is used by the steel industry in the production of high-quality flat steel in the continuous casting process. In the continuous casting operation, a ladle of steel with more than 200 tonnes of molten metal, at 1650°C, is emptied into a 'tundish', a receptacle that holds the steel and controls its flow. Rice husk ash, which is an excellent insulator, with its low thermal conductivity, high melting point, low bulk density and high porosity is used as powder to coat the tundish. It insulates the tundish and prevents the rapid cooling of the molten steel to ensure uniform solidification.

About 0.5 to 0.7 kg of ash is used per tonne of the steel produced. Although this seems to be a minor use, it has been used long enough in many countries, particularly in Europe, to be recognised as an industrial usage. However, in the recent times its use is being questioned on the health safety grounds. It was found that heating of the steel for 4 hours at 1500°C had transformed the silica from its amorphous form to crystalline form with serious health associated risks for the workers.

Cement and concrete

Rice husk ash has been used as a replacement of silica fume, which is a two to four times costlier product, in the production of high performance concrete. It contributes to its high workability, high durability and high strength. Silica fume is also an amorphous form of silica, obtained as a byproduct of burning purified quartz, coal or wood at high temperatures in electric furnace for the production of silicon metal or ferrosilicon alloys. The smoke emanated in the process is collected as silica fume. It is added at about 10% level in the high-performance concrete mix.

Rice husk ash, having low carbon content, could be ground and added up to 20% level in Portland cement to reduce the cost of the cement without compromising on its performance. In fact, it improves the strength of the Portland cement and its resistance to corrosion. Such use not only extends the availability of cement, it also contributes to a corresponding potential decrease in the emission of CO_2 in the atmosphere by the cement industry. RHA has also been used as a replacement for Portland cement in the manufacture of low-strength concrete blocks needed for some special masonry works. Portland cement is not suitable for such requirements.

Refractory bricks

Refractory bricks are used in furnaces which are exposed to extreme temperatures such as blast furnaces used for producing molten iron, and in the production of cement clinker.

Rice husk ash is used in the manufacture of such refractory bricks on account of its insulation properties. With their low thermal conductivity, the bricks made from RHA offer good heat insulation at high temperatures of up to 1450°C. Rice husk ash (80–90%) is mixed with calcium oxide and magnesium oxide (2–20%) as the major constituents for this purpose. However, market for such use is rather small.

Sodium silicate and precipitated silica

Another important area in which RHA has been used, and for which there is an increasing trend, is in the production of precipitated silica and sodium silicate (water glass or liquid glass). These silica-based products have numerous industrial applications. Precipitated silica is classified as one of the 'specialty silicas' along with the other forms, viz., fumed silica, silica gel and silica sol. With an annual growth rate of 5.6%, their world market which was worth US $ 4.5 billion in 2011, is projected to reach US $ 6.4 billion by 2016, with production of 2.8 million tonnes. Precipitated silica accounts for two-thirds of this market. Precipitated silica and fumed silica have high requirements in the rubber industry. The former has a major share in the manufacture of tyres for vehicles, while the latter is used in the non-tyre rubbers. As the level of tyre production is increasing fast, the rubber market is also showing a rise, and so is the demand for precipitated silica. In India the growth rate is even faster, pegged presently at 10%.

In the normal commercial process for the production of precipitated silica, pure quartz sand and sodium carbonate are fused at 1100–1200°C to produce amorphous solid glass (sodium silicate), emitting CO_2 in to the atmosphere. The solid sodium silicate is dissolved in water to give sodium silicate solution (also called water glass), from which pure amorphous silica is precipitated (hence the name) by addition of mineral acid or passing CO_2. Liquid sodium silicate itself is an industrial product and marketed as such for use in many industries, such as detergents, paper, textile, wastewater treatment, adhesive, and construction materials.

Processes have recently been developed using RHA to produce precipitated silica where energy-intensive steps and CO_2 production are avoided and therefore fall under the category of green technologies. However, to be on a good economic footing, the plants should have sufficient continuous supply of RHA. A large plant has been working at Stuttgart, Arkansas, USA, since 2006 that uses 12,000 tonnes of RHA annually to produce 5000 tonnes of precipitated silica and 25,000 tonnes of liquid sodium silicate, and 5,000 tonnes of activated carbon. The firm that set up this plant has now entered into a joint venture with a firm from Sri Lanka to establish husk-fired power plants, and such silica producing plants, in several provinces in Thailand. Processes have been developed in India also, and a firm from Kolkata, has established four RHA-based plants producing precipitated silica in the country.

The demand for precipitated silica got a further boost in the North America recently (which already consumes nearly half of the world produce), with the adoption of 'Green Tires' philosophy. Reinforcement of rubber with precipitated silica is claimed to lower the rolling resistance in tyres, increase the wear resistance and improve wet grip, thus improving fuel economy of the vehicle.

Apart from use in rubber, precipitated silica, with its high surface area (50–300 square metres per gram) and high reactivity, has wide application in many other industries such as plastics, paints and lacquers, cosmetics, ink, pesticides, tooth paste, chemical catalysts, thermal insulation, water softening and desiccants. Its demand and usage, therefore, seems to be growing.

Other potential applications

Minor uses of RHA and its potential for varied application have been demonstrated in many published reports. The Indian Space Research Organization has developed a process for the extraction of highly pure silica which can be used in the manufacture of silicon chips. Attempts have been made to utilize RHA in vulcanizing rubber. In this case, RHA is used as filler in natural rubber compounds to increase its mechanical properties like tensile strength, tear strength, resilience and hardness. Another suggested use is for de-sulpherisation of flue gases from industries, as it has been shown to have a high capacity for the absorption of SO_2 gas.

10.5   Conclusion

The major traditional use of rice husk has been as fuel. This continues to be so even in the present scenario. More than three-fourth of the husk produced is estimated to be utilised for this purpose. Much advancement has taken place over the past couple of decades in the technology for efficient burning of husk and utilisation of its thermal energy in various ways. Cogeneration, based on husk as fuel, is becoming a popular phenomenon. This is moving the large-scale rice processing industry towards self-sufficiency with respect to energy requirements. Systems to produce 50 kW to 30 MW electrical power are operating around the globe. Utilisation of husk for non-fuel purposes, although many in number and for varied applications, forms a minor component. Many potential uses have also been identified, but have remained at exploratory stage and are yet to be tapped for commercial use.

Burning of husk as fuel leaves ash as the residue byproduct. On account of the higher content of silica in it and its porous structure, ash is being utilised efficiently for various purposes, mainly in the construction material and rubber industries. The amorphous form of silica, naturally present in the husk, is retained by burning it below 700°C. It has better prospects for varied applications. Recovered as precipitated silica, its increasing use is seen in a number of industries like rubber, detergents, cosmetics, toothpaste etc.

Further reading

Structure and properties of rice husk and its use have been extensively discussed in:

- Houston (1972)
- Juliano (1985)
- Pillaiyar (1988)
- Luh (1991)
- Champagne (2004)

Use of rice husk for energy purposes has been extensively discussed in:

- Winrock International (1990)

A1.1   The background

Rice mills are often required to purchase and store paddy for their entire year's operations or part thereof as the crop season begins. For example, a large number of rice mills in north India (especially in Haryana and Punjab states) are specifically engaged in milling, processing and trading of basmati rice. The lion's share of their business is in export of basmati rice mainly to West Asia. These enterprises handle some 50–200 thousand metric tonnes of basmati rice in a year each. There is intense competition among the companies. So they generally purchase the bulk of their annual requirement of basmati paddy at the beginning of the paddy season (September to December, depending on the variety) every year.

These procurements evidently run into a very heavy investment and so can be a matter of huge business risk. On the one hand, it requires quick conditioning of the material and its appropriate storage for up to a year under conditions that ensure its safety and protection from deterioration. On the other, it also involves another crucial aspect, namely ensuring the proper quality of the procured paddy itself. Since the paddy is purchased in bulk within a short period soon after the harvest, there is no scope of adjusting the desired quality of the purchase with experience with the progress of the season. The quality of the material has to be assessed and ensured then and there. Similar situations should exist with respect to business in jasmine and other high-quality export rice in Thailand and in other countries in south-eastern and eastern Asia and elsewhere.

It is true that other rice mills elsewhere in India and in other countries (i.e. other than those dealing with basmati, jasmine and such other special rice) may not generally purchase their year's requirement straight away at the beginning of the season. By and large, they may resort to procurement round the year as per their current need. Further, a number of these other rice mills do the bulk of their business by custom milling of paddy procured and owned by the government or other agencies, and so hardly need to purchase much paddy. There is another aspect. Unlike enterprises doing business in basmati or other speciality rice, whose horizon of varieties is limited within a small range, other enterprises deal with a fairly large number of varieties based on demands as currently prevailing in the market. The relative unpredictability and multiplicity of the varieties that thereby result are another reason why these enterprises would hardly need to procure a very large amount of raw material at the beginning of the season. Nevertheless, even these rice mills do procure a substantial amount of paddy to start the season with.

A huge investment is thus required to be undertaken by these mills in a short time at the very beginning of the paddy season. And note that this is without the benefit of a due process of steady and well-considered process of progressive evaluation. It thus becomes a matter of great concern for the enterprises. The situation is generally met by using various rules of thumb to cope with the requirement. These include engaging brokers/agents with long experience for quality assessment and purchase. Some evaluation by in-house experts of quality based on their own rules of thumb is also done. These criteria may include assessment of grain colour, its gloss, shape, plumpness, health, incidence of infestation or mould damage, and grain breakability as perceived by rubbing a small amount of paddy between the palms.

These rules of thumb no doubt do provide a reasonable assessment of the quality of the material. However, their limitations are obvious. Precise and reproducible, yet simple and quick, laboratory methods of assessment of the quality or value of the paddy being purchased are clearly necessary.

A1.2   Aspects of quality to be assessed

There are two aspects of value assessment of the paddy. One is its inherent (genetic) grain quality, viz. the cooking quality, the eating quality, taste, texture, flavour, etc. and perhaps the product-making quality of the rice. These attributes are dependent on the inherent chemical and physicochemical properties (i.e. genetic make-up) of the material and are largely subsumed in the variety. Once the variety and the source are identified and verified, these inherent qualities are by and large automatically selected and more or less ensured. Such quality attributes of the procurement can therefore be more or less excluded from the scope of the initial quality assessment being discussed. Such assessments as required would be done during the working season as the end products get produced. In other words, the main object of value assessment of the raw material being purchased, i.e. the intended procurement, is primarily to check the outturn and wholesomeness of the material including its grain integrity. To put it in a nutshell, what is required is to ensure that one gets the maximum amount of standard quality saleable product of appropriate grade from the grain being procured.

Clearly, there are five aspects to be considered:

(a) First is the amount of impurities (sand, stone, mud balls, dust, chaff, straw, other seeds) in the procured material. Impurities in the grain are a direct loss to the buyer and are therefore to be fully discounted for.

(b) Second is the moisture content of the paddy to be procured. The greater the moisture in the paddy being purchased beyond the safe-storage moisture level (of say 13%), the lesser obviously would be the saleable rice that can be produced from it.

(c) Third is the total milling yield (or, more generally, total brown rice yield) of the paddy. After all, ultimately it is not the paddy but the inner rice grain that the miller is interested in. So, the more the brown or milled rice output (or the less the husk content), the more is the value of the procurement.

(d) Fourth is the amount of breakage (broken grains or its obverse, the head rice) within the brown (or milled) rice outturn. Rice is unique among cereals in that it is overwhelmingly being used in the whole-grain form. Any breakage of the grain during its milling is therefore a direct loss to the producer. So grain breakage or the head-rice yield of the procured material is perhaps its most important attribute that has to be assessed before purchase.

(e) Finally, there may be various refractions that the paddy contains. These are red grain, chalky grain, heat-damaged grain, green grain, immature grain, etc., that pull down the market grade of the product. These can be dealt with in one of two ways. In some countries strict limits for these refractions are prescribed and various grades are prescribed for different ranges of these refractions. This is true for example of USA. In such cases, the amount of refractions has to be ascertained so as to assign the proper grade to the material being procured. In other countries, specific discount rates for the refractions, instead of grade assignment, may be involved. In such situations, precise assessment of the quantities of these refractions becomes necessary.

It thus becomes clear that ideally for annual procurement of paddy, laboratory procedures need to be developed and put in place which can quickly yet precisely verify the amounts of

- moisture content
- brown-rice yield
- grain breakage and/or head brown-rice yield and
- various refractions, viz. red grain, chalky grain, heat-damaged grain, green grain, immature grain, etc.

It may look surprising that the first item listed in the earlier paragraph, viz. the content of impurities in the procured paddy, is missing in the present list of precise items to be tested. Actually, this is not a surprise. If the sample of the paddy under test is directly shelled, i.e. dehusked (after necessary drying but without cleaning), and if the quantity of outturn of saleable products is estimated directly on that basis, then the loss in outturn caused by the presence of any external non-grain material (impurities) would be automatically taken care of. These impurities would not obviously contribute any brown rice upon shelling, so their negative effect would be automatically contained in the dehusking result. Thus there is no essential requirement of separately estimating the amount of impurities in the material. However, if the paddy contained significant amounts of stone and sand which might cause damage to the paddy sheller, then the situation might be different. In such a case, it would be better to estimate the content of impurities separately, carry out all the tests with cleaned paddy, and then introduce a correction factor into the results to account for the amount of impurities.

A1.3   Outline of the laboratory tests

The following few simple laboratory procedures or analyses would enable one to determine the criteria mentioned above.

(1) Precise sampling of the grain under procurement is the first requirement. This has to be done by following the procedures listed and explained in standard laboratory hand books. Such sampling should not only include collection of small samples from different stocks (including bags), but also the use of an appropriate Boerner divider or other sample reducer to reduce the original sample to the test quantity. The appropriate sample is then used for testing the criteria below.

(2) The first job is to estimate the moisture content of the paddy. This should be tested in a subsample originally obtained during the procurement and carefully preserved in an appropriate plastic package. The moisture estimation of the large number of samples would necessarily have to be done by a moisture meter. The latter must therefore should have been previously carefully calibrated against a standard oven procedure. This last aspect is discussed in Appendix 2.

(3) The next task is to dry the test sample to approximately 13% moisture so as to make it suitable for laboratory analysis. The idea here is not to dry the sample to a precise moisture content nor is it necessary to know the precise amount of moisture in the dried

sample. All that is needed is to dry all the test samples more or less uniformly to near about 13% moisture so as to make them suitable for shelling and further analysis. The exact moisture content has no bearing on the results as long as it is close to 13%. It is best to equip the laboratory with a multi-sample laboratory warm-air dryer for the above purpose (such as the Satake Tray Dryer, which can handle 48 samples at a time). Given no pressure of time, the drying could be performed by various means. But since a continuous stream of incoming samples is expected to arrive for analysis, the laboratory ought to be equipped with a multi-sample warm-air (air temperature ≤40°C) dryer. It should be provided with suitable controls such that individual channels can be switched off or be automatically stopped at the time set when the sample is expected to reach around 13% moisture. The drying should not be performed in one stretch if the moisture content is rather high (say >15%), but in two stages, with a rest in between. The dried sample should then be stored in a plastic package for at least 12 hours before it is taken for further analysis.

(4) The sample may or may not now (or before the drying step) be passed through a suitable dockage tester. This step may not be necessary in the absence of significant quantities of sand and stone in the paddy. However, if the analysis requires the quantity of immature grain contained by the sample to be quantified (see below), then a dockage tester may have to be used. In that case one can also determine the contents of (i) impurity, (ii) immature grains and (iii) other seeds, etc. in the sample during the same test.

(5) A weighed sample of the dried paddy (say 300–500 g) should now be shelled (dehusked) in a laboratory sheller (preferably a rubber-roll sheller, such as the Satake laboratory sample sheller model THU). The sheller should be so adjusted that approximately 85% of the grains is shelled in a pass, leaving around 15% of the paddy grains unshelled. Complete or near-complete shelling in a single pass would cause excessive grain breakage and hence provide incorrect results. The total quantity of the outturn should be carefully weighed. The output is now passed through a sizing device or an indented cylinder to separate the broken and head brown rice (and unshelled paddy) grains. Ideally the unshelled paddy grains should be separated (using a dockage tester or manually), shelled separately and this output combined with

the previous brown rice output. But in routine testing of a large number of samples in the procurement season, such elaborate procedure is just not possible. So the percentage of undehusked grains in a small subsample is estimated, this amount is subtracted from the amount of paddy taken for dehusking, and the outturn of brown rice and various fractions are calculated on that basis (i.e. on the basis of the amount of paddy actually shelled).

(6) Then, an accurately weighed subsample of the sheller output (10–50 g) is to be taken for analysis of (a) unshelled paddy, (b) head rice grain, as well as (c) various refractions within the head brown rice grain (red rice, chalky rice, damaged and discoloured grains, etc. mentioned earlier).

These data are then used in the formula below to arrive at a measure of the value (value quotient) of the paddy.

A1.4   The value quotient formula

The formula of the value quotient (VQ) of the paddy is based on the following assumptions:

(a) Moisture of the final products (i.e. saleable brown or milled rice) is assumed to be 13% (wet basis). Any excess moisture over 13% in the paddy is therefore discounted for.

Note – The value of 13% is purely arbitrary. Any other value (12, 14, 16, even zero) can be taken. But then, that value has to be substituted for 13 in the formula below.

(b) The ideal yield of brown rice from paddy is taken as 78%.

Note – This 78% again is purely arbitrary. Any other value (75, 80 or any other) can be taken. But then, that value should be used to replace the value of 78 in the formula below.

(c) The entire VQ analysis is based on the outturn of brown rice, rather than white rice. This is done for the sake of convenience in view of the likely large number of analyses to be carried out every day during the procurement season. An analysis based on further milling (whitening or debraning) of the brown rice would be difficult to cope with.

(d) The products of shelling are arbitrarily valued or discounted for as illustrated below. These valuation rates are purely arbitrary and can be replaced by other sets of values. The various items, their symbols and their valuation/discount rates (arbitrary) are as follows:

Item	Symbol	Proportion	Valued as
Moisture of paddy at purchase	M	% of paddy (wet basis)	–
Head brown rice	HR	% of paddy	1.0 (i.e. 100%)
Broken brown rice	Br	% of paddy	0.3 (i.e. 30% of head rice)*
Chalky head grains	Ch	% of HR	0.5 (i.e. discounted by 50%)*
Green head grains	Gr	% of HR	0.7 (i.e. discounted by 30%)*
Damaged and discoloured head grains	DD	% of HR	0.2 (i.e. discounted by 80%)*
Red head grains	R	% of HR	0.3% (i.e. discounted by 70%)*
Immature grains in head rice	Im	% of HR	0.2 (i.e. discounted by 80%)*
Immature grains in paddy	Im-P	% of paddy	–

*These valuations are purely arbitrary and are shown only for illustration. Then the value quotient (VQ) is:

$$VQ = \frac{HR + 0.3\,BR - \dfrac{HR}{100} \times 10.5\,Ch + 0.3\,Gr + 0.8\,DD + 0.7\,R + 0.8\,Im}{78} \times 100 \times \frac{100 - M}{87}\,\%$$

It may be noted that the above formula gives an index of the value of the sample in comparison to an ideal sample of paddy. That is, if the ideal sample is considered to have 100% value, the value quotient VQ would indicate what the value of the test paddy is as a percentage of that of the ideal. An ideal paddy may be assumed to give 78% brown rice yield after dehusking, all head rice and no broken grains, no refractions (chalky grain, immature grains, etc.), and 13% moisture. Then by substituting these values in the above formulae, the VQ would come out as 100%. Other samples would give an index less than 100% depending on the contents of various items. Finally, the value as obtained on the basis of various outputs of rice and refractions is appropriately discounted by the amount of excess (or less) moisture over 13% (wb) in the procured paddy to arrive at the final VQ. As explained in Chapter 3 under expressions of moisture content, the correction can be done not on wet basis moisture but only on dry basis, that is, only on the basis of the solid matter. The solid matter in the procured paddy is $(100 - M)$. In the ideal paddy it is $(100 - 13 = 87)$. That is how the last item in the formula on right is used for the moisture correction.

A1.4.1   A possible error

An error may crop up in the calculation of value quotient as above on account of too many immature rice present in the sample. As discussed under the test procedure, laboratory shelling of paddy is a tricky job. In the regular production in the rice mill, the sheller is set to shell around 80–85% of the paddy grains in one pass to minimise grain breakage. Complete (100%) dehusking in a single pass is never advisable, for that would cause excessive grain breakage. Besides, there may be some relatively thinner grains in the paddy (which may tend to be immature or otherwise somewhat fragile). Trying to get them all dehusked in one pass may cause excessive stress on the normal grains, leading them to tend to break. That is why only 80–85% grains are shelled in the first sheller pass. The product of the sheller is now fractionated in the paddy separator into dehusked brown rice and undehusked paddy, the latter being returned to the sheller (or a second sheller) for reshelling. Ideally, the same procedure should be done in laboratory testing. But in practice, this is hardly ever possible. For one thing, there are no simple laboratory equipments available for paddy separation, therefore complete separation of paddy and rice (for reshelling of the undehusked grains) is an arduous task. There are ways to meet the situation for careful and accurate work (see details in Bhattacharya 2011). But this can hardly be adopted for routine tests, leaving one to do the best that one can in a shortcut. In normal testing where perhaps 1 kg paddy is taken for testing and the broken and head rice are got separated by a sizing device, the undehusked paddy would go along with the head rice. It is then not very difficult to separate most of the paddy from the mixture manually or with the help of a sizing tray and get it reshelled. But such is not possible where a large number of samples are to be mass tested.

The error comes from the immature paddy in certain samples, which are rather fragile and liable to break during shelling or milling. Now most varieties or samples hardly ever contain appreciable amounts of such immature paddy grains. Not much difficulty arises there. But the problem arises in a few varieties where the samples may have large or variable proportions of fragile immature grains (2–5%). What happens in these cases is that most of the immature grains escape shelling, so that the unshelled grains in the output of the dehusker are all largely immature grains that would have otherwise yielded broken rice if shelled. If this unshelled paddy is now deducted from the amount of test paddy to calculate the percentage of various fractions, the result would be vitiated by that amount of potential broken rice being excluded. Therefore by escaping being shelled at the dehusking stage, and then getting deducted from the paddy input, they lead to an erroneously low result of broken grains and hence an artificially inflated VQ value.

The best that can be done in such cases is as follows. Separation of paddy grains and reshelling them for all samples in the procurement season is impossible. What may be done is to assess the approximate degree of error that is thus introduced. A few carefully selected samples with low, medium and high contents of immature grains are tested both by the test procedure described above and a second sample with 100% shelling (i.e. about 85% shelling first, followed by manual separation and reshelling of the unshelled grains). Once the relationship between the two sets of VQ values is thus found out, an approximate correction can be introduced into the results. First one should determine the amount of immature paddy in the sample by passing the paddy through a dockage tester. Now a correction can be introduced into the final VQ result on the basis of the immature grain contents thus found and its impact on the VQ determined above.

Appendix 2:

The determination of moisture in paddy/rice using a moisture meter

A2.1   The background

Moisture is an integral part of any biological material, living or dead, including grains. Besides, this moisture is usually not a passive entity but influences the wholesomeness and behaviour of the material. The moisture content of a material often depends on the circumstances around it and changes as the circumstances change. The property of the material at any given time may change depending on such variation in its moisture.

As discussed in Chapter 3 (Drying of Paddy) and Chapter 4 (Storage of Rice), moisture content of rice (whether paddy, brown rice, milled rice or bran) changes with the prevailing humidity of the surrounding air. It also changes due to natural factors. For instance, paddy has a fairly high moisture when it is harvested, which then gets lowered naturally or else by artificial drying. And the specific moisture at any given time may influence the storability or the usability or any other utility of the grain or any other material.

So determining the amount of moisture the grain has at all stages is a major factor in its processing and management. Besides, continuous monitoring of the moisture content often becomes necessary in monitoring a process. For instance, the moisture content needs to be continuously monitored during drying of paddy. Similarly during parboiling of paddy, moisture content of the grain changes continuously during soaking, and later continuously again during drying of the product. Crucial changes in the product and processes often happen at specific stages in its processing. For example, the grain becomes liable to fissure or crack during heated-air drying of wet parboiled paddy as its moisture content drops to 16% or below. So the precise monitoring of this stage for the purpose of determining when to terminate the first stage of its drying is essential during heated-air drying of parboiled paddy. Monitoring of the moisture content is also needed to determine when to terminate the drying process itself (12–14% moisture). The same thing applies to the drying of field paddy. Again the moisture content of the grain is often required to be maintained at specific levels for optimal operation of a process. For example, steam processing of new (i.e. freshly harvested or not yet aged) paddy is best done at approximately 13–14% moisture. Similarly, it is essential to know the exact moisture content of the grain during the procurement season for its correct valuation.

To conclude, the moisture content is a crucial component of rice, as of any other biological material, and its careful monitoring at every stage is often a critical requirement during its various processes. Clearly, quick and reasonably accurate determination of the moisture content is a standard requirement in any rice-processing unit (as of any other grain-processing facility). Obviously, being more or less like an on-line determination, such routine monitoring can only be done by the use of a simple and quick moisture meter. Such moisture meters are widely available. But, to be able to use it profitably, it is essential for one to be knowledgeable about its strengths, weaknesses and hazards.

A2.2   Use of a moisture meter

A2.2.1   How a moisture meter works

The first point one should understand about a moisture meter is that a moisture meter does *not* directly measure the moisture! This may sound absurd to the uninitiated, but it is actually true of most if not all meters. By and large meters read not the property it is intended to measure directly, but actually some other easily measurable effect of that property. This can be illustrated by the case of a glass thermometer. The glass thermometer is a simple instrument consisting of a reservoir for mercury and a capillary tube in which the mercury may rise when it expands. When the thermometer is placed in a hot material (to measure its temperature), the mercury in the reservoir expands due to the heat and so rises in the capillary. When it is put in a cooler material, the mercury contracts, so it comes down in the capillary. So the thermometer actually measures only how much the mercury has expanded or contracted at a particular moment. But this change in the mercury volume is only an effect of a change in temperature. Therefore, if the rise or fall of the mercury in the capillary is marked (calibrated) in terms of the temperature, then the same gets converted into measuring the temperature. In other words, we are only measuring the effect of temperature (mercury volume), and are then converting that measurement into a measure of the temperature by calibrating the mercury volume against the temperature.

A more or less similar thing happens in the moisture meter. In the moisture meter, i.e. in the resistance or conduction type which is most commonly used, a current is made to be passed between two electrodes with the material (such as rice), the moisture of which is to be measured, being placed between the electrodes. So the moisture meter actually measures only the amount of current that flows through the rice between the electrodes. Now it so happens that the amount of current flowing between the electrodes here depends on the moisture content of the material placed between them. The more damp the material is, the more current flows and conversely, the drier the material, the lesser the current flow. So once this relationship between the moisture content and the current is calibrated, the meter can indirectly measure the moisture. Clearly, the proper calibration of the meter is an essential requirement for its use. This situation is actually true of nearly all meters. Usually the meter measures something, but that something is calibrated in terms of something else, i.e. the property in which we are interested. And thus we indirectly measure the property we want.

A2.2.2   Importance of calibration of a moisture meter

Proper calibration of the meter is thus a crucial part in the use of a meter. In the case of a moisture meter, normally the meter should always give a correct measure of the current that flows. But knowing how much moisture that amount of current is equal to is a matter of the calibration. If the calibration is correct, the result will then be quite correct. If the calibration is improper, the indicated moisture reading can be wide off the mark.

So, apart from following the correct operating procedure, the accuracy of the moisture meter is dependent on the accuracy of its calibration. No doubt the manufacturer does carry out a calibration of its own and provides an appropriate chart for converting the reading into the moisture (or the meter scale itself is marked in terms of moisture). However, we should remember that a moisture meter is meant to test a variety of materials (such as paddy, milled rice, wheat, other grains, other milled products, etc., etc.). Therefore, to expect that one or a few calibrations would suffice for a variety of materials is to expect too much. There are many factors which may render such calibration inadequate. Relying blindly on the meter reading ('trust-meter syndrome') is thus a serious hazard in the use of a moisture meter (as of all other meters).

A2.2.3   The process of calibration

As explained, a meter usually reads and expresses the values of a property A (e.g. mercury volume or current flow). These A values (A1, A2, A3,...) are then converted to B values (e.g. temperature or moisture) by calibrating the A values in terms of B values (B1, B2, B3,...). So the meter can now measure property B. This calibration is done by reversing the process of using the meter. That is, instead of using the meter to measure unknown B values of a material, known B values of the material are used to mark the meter, i.e. to mark their corresponding A values in the meter. Once this process is carefully done, the meter is now ready to be used to measure unknown values of property B.

So the calibration of any meter requires the availability of a set of standard samples whose contents of the desired property are already known from some other source. For example, if a sample is previously known to have a moisture content of 20% and if it shows the reading of value 'x' in the meter, then the value 'x' can be marked as 20% moisture. If a series of such standard samples are examined and their readings are noted, these readings now provide a calibration chart.

Let us illustrate the matter by taking the case of a thermometer. A thermometer actually shows expansion or contraction of mercury. This expansion or contraction is caused by change in temperature. So by calibrating the mercury volume against temperature, we can use the thermometer to measure temperature. What we need for that calibration is a few samples of materials having known temperatures. That is, we need materials with reference standard temperatures. Fortunately, in this case we do not require the set of reference standard materials to be procured from elsewhere. The reference standards are freely available to all. This is because Nature herself has provided two standards which we may freely use. Under standard atmospheric conditions, water boils at 100°C and freezes (turns into ice) at 0°C at mean sea level. So by placing a thermometer in boiling water, we can mark the resultant mercury position as 100°C. Similarly by placing the thermometer in a mixture of ice and water we can mark the position of mercury shown as 0°C. By dividing the intervening length properly, we have a meter for measuring temperature.

Let us now compare this with a moisture meter. As we noted, the amount of current that flows between the two electrodes depends on the moisture content of the material placed between the electrodes. Therefore, if we have a series of standard samples having known but different moisture contents, then by testing these materials in the meter, we can prepare the calibration chart of the meter reading against the moisture.

So the task here is two-fold. The first task is to prepare a set of standard samples of the material (paddy, milled rice or whatever is to be tested) having different amounts of moisture, say in the range of 10–25%. This can be done, for instance, by adding different appropriate amounts of water to different samples of the material (e.g. paddy). The water is mixed well and then the paddy can be put in tight plastic pouches and left overnight to equilibrate. These samples can then be tested next day both by a standard procedure (to know its true moisture) and by the moisture meter (to calibrate it).

The second task is to determine the actual moisture contents of these samples by a standard procedure. For this, one has to normally depend on a standard oven-drying procedure. The assumption in this procedure is that if the material is heated under certain standard conditions, it would lose all its moisture and moisture alone (nothing else). Thereby the loss in weight undergone by the material when heated under the prescribed standard conditions is considered as equal to its moisture content. One can thus prepare a set of samples of the material having varying moisture contents, as described above. The true moisture of these samples are then determined by the above standard oven procedure. This set is now tested in the moisture meter and the corresponding points are noted or marked. One can in this way now have an appropriate calibration chart with which to determine the moisture content of unknown samples of the same material. Such standard oven procedures are described in various standard books of analysis and can be obtained there from. In the absence of an in-house laboratory, the same system of calibrating the moisture meter with the help of a nearby Central Laboratory would have to be followed. In the minimum, a set of standard-moisture samples should be procured from a consulting laboratory and use them to calibrate the house meter.

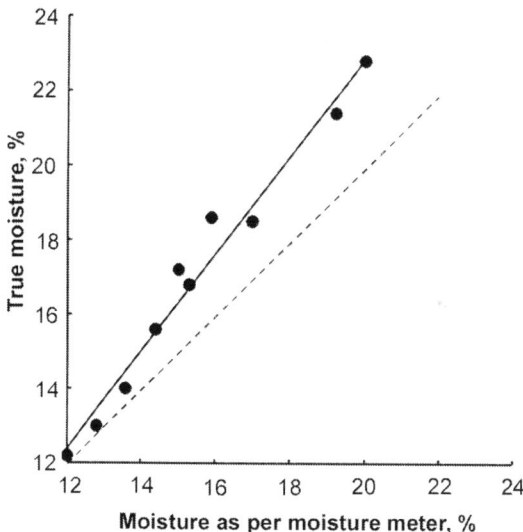

Figure A2.1 A moisture-meter calibration curve. Had the meter shown the true moisture (as indicated by a standard oven-drying method), the curve would lie close to the 45° dashed line [Reproduced with permission from Bhattacharya (2011)]

A real-life calibration of a moisture meter carried out by the author in a rice mill is shown in Fig. A2.1. The true moisture contents of the samples of paddy as determined by a standard oven-drying method were plotted against the moisture values as indicated by the moisture meter. The discrepancy between the two sets of data is obvious. For example, when the meter showed a moisture content of 14%, the true value was close to 15%. So if, say, the lot was undergoing drying and the drying was terminated at this point thinking that the safe-storage moisture level had been reached, the lot would very likely spoil later because of the high actual moisture (15%). Note further that the difference between the two sets of values went on increasing as the moisture content increased. So one cannot simply add a fixed correction factor (e.g. 1% in the above example). For instance, we have seen in Chapter 7 (Parboiling of Rice) that during drying of parboiled paddy, the drying must be halted and the paddy allowed to temper at as close to 16% moisture as possible. It can be seen from the Figure that when the meter showed 16% moisture the true moisture was nearly 17.5%. So, if the operator went by the meter value, the tempering would occur too early, making the lot liable to grain-fissuring during its second drying stage. Note further that an additional correction would be required in the example cited, in view of the paddy being under drying. This is explained in Section A2.3 below.

A2.2.4   Need for separate calibration for each material

While the above process should suffice to use a moisture meter accurately to determine the moisture of one material, it may still be not sufficient for use with different materials. This matter will be better understood by again comparing it with the case of a thermometer. The expansion and contraction of the volume of mercury in the thermometer depends on the temperature of the test material, and temperature alone. It is not influenced by any other property of the test material. So we can confidently use it to measure temperature of diverse materials. But that is generally not the case with other material properties. For instance let us take this case of the amount of current that flows through a material (e.g. paddy) placed between two electrodes. This is no doubt dependent on the paddy moisture. But it depends on other properties of the paddy as well – e.g. its composition, especially its content and composition of minerals. As this composition is liable to vary from material to material, the effect on the current due to factors other than due to moisture is also liable to vary. Now a cardinal principle of any measurement is that the property being measured should depend on one property (or variable) only. If it depends on more than one variables, then the effect cannot be used to measure either of the variables directly. It can be used only after nullifying the effect of all the other variables and leaving only one that is to be measured.

So, after proper calibration with a material (e.g. paddy), as explained, a moisture meter can be used to measure the moisture content of that material (i.e. paddy) alone. If it has to be used to test other materials (brown rice, milled rice, wheat, flour, sorghum, etc...), then a fresh calibration with that material should be carried out. In other words, for the proper use of a moisture meter, it should be separately calibrated for each material that is intended to be tested by it. Raw and parboiled rice too should be calibrated separately. Sometimes even different varieties of a given material may differ in their response and so have to be calibrated separately.

A2.3   Moisture content of material undergoing drying

The problem of using a moisture meter in grain-processing enterprises unfortunately does not end here. Even after proper multiple calibrations have been prepared, the meter may still fail to provide the true moisture content of a material. This happens particularly when the material under test (for example, parboiled paddy) is undergoing drying or soon after it has been dried. Why this happens and the remedy thereof can be understood from the following discussion.

There are two phenomena involved here. First is the behaviour of the moisture meter or current flow. When the moisture of a particulate or whole grain is tested, the meter generally senses the moisture content of the grain's surface layers rather than of its core. The reason this happens is probably that the mineral constituents of the grain are mostly located in the grain's surface layers. And the minerals are the better conductors of current in the grain. Be that as it may, if the moisture in the grain is not uniformly distributed due to any reason, the moisture meter will mostly record the moisture in the surface layers only and not the aggregate average moisture. If the grain is first ground into a powder before testing, then all the grain layers get well mixed and the aggregate value is shown. But the moisture meters that are manufactured are generally designed to use whole grains in the procedure and then this problem arises.

Why the moisture content may differ in different layers of the grain during its drying has been discussed in Chapter 3 (Drying of Paddy). The problem arises because of a second phenomenon with materials undergoing drying. When a material such as grain undergoes drying, moisture evaporates only from its surface and not from the inner layers. The moisture inside migrates from the centre to the surface by a process called diffusion and only then can it evaporate. Now diffusion is a much slower process than evaporation. So the surface dries out faster than the inside. The end result is that whenever a grain is undergoing drying, a moisture gradient develops within it. The surface of the grain becomes drier than the inner matter and the inside remains more wet than the surface. As the surface is dry and, as stated before, the meter reads the moisture only of the grain surface, the meter reports a reading of moisture which is much less than the aggregate average moisture. For example, if the aggregate average moisture content is 16%, the central core of the grain may have a moisture content of 20–25% and the surface layer may have a moisture content of 10–12%. If this grain is now tested in a moisture meter, it will, as explained above, read only the moisture of the surface layer and therefore show the grain moisture as only 10–12%.

Interestingly if the grain above is now kept in a closed plastic pouch overnight and then tested again the next day, it will now show the correct reading of 16%. This is because the grain moisture became equalised within the grain during the overnight rest. While the evaporation from the grain surface had stopped upon resting, the diffusion from wet to dry region continued until the difference between layers disappeared. The operator thinks and generally reports that the material absorbed some moisture during the night! This is a very common misconception. But that perception is of course wrong. Moisture remained the same as before. Earlier (during or soon after drying) it was unevenly distributed and the meter correctly read 10–12% (in the surface). After resting, the moisture became equalised throughout, and the meter again correctly read (again in the surface) 16%. So the meter was not wrong. It was giving the correct result both the times. The fault was ours that we did not understand the matter.

The remedy for this second problem is also the same, viz. again calibration. In other words, the moisture meter is to be calibrated not once but twice. The first calibration is for the moisture in general, that is in the rested material. A second calibration is needed to find out the reading in the grain undergoing drying or soon after drying. The procedure is the same. One should collect samples at different stages of drying. These samples should be tested in the meter twice – once immediately and once again after overnight storage in a sealed plastic pouch. The difference between the two readings is the correction factor to be added to the reading when testing material during or soon after drying.

The importance of the matter can be illustrated by going back to the same example cited above (in the concluding part of Section A2.2.3). It was mentioned that tempering of the parboiled paddy undergoing drying (intended to be done at 16% moisture) would be at an improper stage if decided from the reading of an uncalibrated moisture meter. But this was only a part of the story. The actual error would be still more. As the paddy was undergoing drying, the meter reading would have an additional error as explained here. As a result, when the meter showed 16% moisture, the true aggregate grain moisture would be not 17.5% as mentioned above but perhaps 20% or more, rendering the whole process useless.

A2.4   Conclusion

This discussion and the procedures mentioned may at first sight look a bit complicated and rather difficult to follow. But the matter is actually very simple. Once the basic principles are understood, the matter would turn out to be simple and accuracy can be ensured with ease. The moisture meter will then become a simple but valuable and indispensable tool in the hands of the enterprise.

A moisture meter, like all meters, can be an essential and invaluable tool as long as it is treated like a servant (service agent) under the control of the operator as the master. If it is treated like a master controlling the operator ('trust-meter syndrome'), then it can cause serious damage to the enterprise.

Further reading

Standard methods for estimation of moisture in food or other biological materials have been well described in all analytical handbooks. Use of moisture meter for determination of moisture in paddy and rice has been dealt with in detail in Bhattacharya (2011).

References

ALI S Z and BHATTACHARYA K R (1980), 'High-temperature drying-cum-parboiling of paddy', *J Food Process Engg*, **4**, 123–136.

ARAULLO E V, DE PADUA D B and GRAHAM M (1976), *Rice Postharvest Technology*, Ottawa, Canada. International Development Research Centre.

BAILEY C H (1940), 'Respiration of cereal grains and flaxseed', *Plant physiol*, **15**, 257–274.

BARBER S (1972), 'Milled rice and changes during aging', in Houston D F (Ed.), *Rice Chemistry and Technology*, St. Paul, MN, AACC International, 215–263.

BARBER S and BENEDITO DE BARBER C, *Rice Bran: An Under-Utilized Raw Material*, New York, NY, UNIDO, United Nations, 251.

BHATTACHARYA K R (1985), 'Parboiling of rice', in Juliano B O (Ed.), *Rice: Chemistry and Technology*, 2nd edn, St. Paul, MN, AACC International, 289–348.

BHATTACHARYA K R (2004), 'Parboiling of rice', in Champagne E T (Ed.), *Rice: Chemistry and Technology*, 3rd edn, St. Paul, MN, AACC International, 329–404.

BHATTACHARYA K R (2011), *Rice Quality: A Guide to Rice Properties and Analysis*, Cambridge, England, Woodhead Publishing, 578.

BHATTACHARYA K R (2013a), 'Process for accelerated aging of rice', *Cereal Foods World*, **58**: 19–22.

BHATTACHARYA K R (2013b), 'Improvements in technology of parboiling rice', *Cereal Foods World*, **58**: 23–26.

BHATTACHARYA K R and ALI S Z (1970), 'Improvement in commercial sun drying of parboiled paddy for better milling quality', *Rice J* **73**(9): 3, 4, 9, 12–15.

BHATTACHARYA K R and ALI S Z (1985), 'Changes in rice during parboiling, and properties of parboiled rice', in Pomeranz Y (Ed.), *Advances in Cereal Science and Technology*, Vol. VII, St. Paul, MN, AACC International, 105–167.

BHATTACHARYA K R and SUBBA RAO P V (1966), 'Processing conditions and milling yields in parboiling of rice', *J Agric Food Chem.* **14**, 473–475.

BHATTACHARYA K R, ALI S Z and INDUDHARASWAMY Y M (1971), 'Commercial drying of parboiled paddy with LSU dryers', *J Food Sci. Technol.*, **8**, 57–63.

BOND N (2004), 'Rice milling', in Champagne E T (Ed.), *Rice: Chemistry and Technology*, 2nd edn, St. Paul, MN, USA, 283–300.

BORASIO L and GARIBOLDI F (1957), *Illustrated Glossary of rice Processing Machines,* FAO Agric Serv Bull 37, Rome, Italy, Food and Agric Org of the UN.

CHAMPAGNE E T (Ed.) (2004), *Rice: Chemistry and Technology,* 3rd edn, St. Paul, MN, AACC International, 640.

CHEN SAN FUNG MACHINERY (2014), 'Rice Polishing machinery (Friction

Type), Product Brochure, Tantzu, Taichung, Taiwan (http://www.csfm.com.tw).

CHRISTENSEN M (Ed.) (1982), *Storage of Cereal Grains and their Products,* 3rd edn, St. Paul, MN, AACC International.

DACHTLER W C (1959), *Research on Conditioning and Storage of Rough and Milled Rice,* ARS 20-7, Agric. Res. Service, US Dept. Agric, 55.

DRAYTON J (1802), A View of South Carolina as Respects Her Natural and Civil Concerns, Charleston, SC, USA, W P Young.

ESMAY M, SOEMANGAT, ERIYATNO and PHILLIPS A (1979), *Rice Post-Production Technology in the Tropics,* Honolulu, East-West Center, Univ. Hawaii, 140.

FABIAN C and JU Y-H (2011), 'A review on rice bran protein: Its properties and extraction methods', *Crit. Rev Food Sci Nutri,* **51,** 816–827.

FOSTER G H (1982), 'Drying cereal grains' in Christensen C M (Ed.), *Storage of Cereal Grains and their Products,* 3rd edn, St. Paul, MN, AACC International, 79–116.

FRIEDMAN M (2013), 'Rice brans, rice bran oils, and rice hulls: Composition, food and industrial uses, and bioactives in humans, animals, and cells', *J Agric Food Chem,* **61,** 10626–10641.

FULLER D Q (2011), 'Pathways to Asian civilizations: Tracing the origin and spread

of rice and rice culture', Rice **4,** 78–92.

GARIBOLDI F (1974), *Rice Milling Equipment Operation and Maintenance,* FAO Agric Serv Bull 22, Rome, Italy, Food and Agric Org of the UN.

GARIBOLDI F (1984), Rice Parboiling, FAO Agric Serv Bull 56, Rome, Food Agric Org of the UN.

GHOSE R L M, GHATGE M B and SUBRAHMANYAN V (1960), *Rice in India,* Rev. edn, New Delhi, India, Indian Council Agric Res.

GLASZMANN J C (1987), 'Isozymes and classification of Asian rice varieties', Theor Appl Genet, 74, 21–30.

GRiSP (2013) (GOBAL RICE SCIENCE PARTNERSHIP), *Rice Almanac,* 4th edn, Los Banos, Philippines, International Rice Research Institute.

HOUSTON D F (Ed.) (1972a), Rice: Chemistry and Technology, St. Paul, MN, AACC International.

HOUSTON D F (1972b), 'Rice hulls' in Houston D F (Ed.), *Rice: Chemistry and Technology,* St. Paul, MN, USA, AACC International, 301–352.

HOOGENKAMP H (2012), Rice Bran Protein, Henk Hoogenkamp, www.henkhoogenkamp.com.

IRRI (2007), *Annual Report,* Los Ban☐os, Laguna, Philippines, International Rice Research Institute.

ISHII I (1995), 'Rice mills and rice milling systems of Yamamoto Manufacturing Co.', in Hosakawa A (Gen. Ed.), Rice Post-Harvest Technology, Tokyo, Japan, The Food Agency, Ministry of Agriculture, Forestry and Fisheries, 406–415.

JULIANO B O (Ed.) (1985), Rice: Chemistry and Technology, 2nd edn, St. Paul, MN, AACC International, 774.

JULIANO B O (1995), 'Rice hull and straw', in Juliano B O (Ed.), Rice: Chemistry and Technology, 2nd edn, St. Paul, MN, USA, 689–755.

JULIANO B O (2007), *Rice: Chemistry and Quality,* 2nd edn, Los Ban☐os, Laguna, Philippines, Philippines Rice Research Institute.

KATSURAGI Y (1995), 'Rice milling machines', in Hosakawa A (Gen. Ed.), *Rice*

Post- Harvest Technology, Tokyo, Japan, The Food Agency, Ministry of Agriculture, Forestry and Fisheries, 351–361.

KUNZE O R and CALDERWOOD D L (2004), 'Rough-rice drying–Moisture adsorption and desorption', in Champagne E T (Ed.), *Rice: Chemistry and Technology,* 3rd edn, St. Paul, MN, AACC International, 223–268.

KUNZE O R, LAN Y and WRATTEN F (2004), 'Physical and mechanical properties

of rice', in Champagne E T (Ed.), *Rice Chemistry and Technology,* 3rd edn, St. Paul, MN, AACC International, 191–221.

LEVEE A P (1940), *The Development of Rice Industry in the Lake Charles District,* M.A. thesis, Univ. Texas, Austin, Texas, USA.

(http://ereserves.mcneese.edu,/depts/archives/FT/Books/Levee.html)

LU J J and CHANG T–T (1980), 'Rice in its temporal and spatial perspective', in Luh B S (Ed.), *Rice: Production and Utilization,* Westport, Connecticut, USA, The Avi Publishing Co. Inc., 1–74.

LUH B S (1980), *Rice: Production and Utilization,* Westport, Connecticut, Avi.

LUH B S (1991a), *Rice: Production,* vol. 1, 2nd edn, New York, NY, Van Nostrand Reinhold.

LUH B S (1991b), *Rice: Utilization,* vol. 2, 2nd edn, New York, NY, Van Nostrand Reinhold.

MACLEAN J L, DAWE D C, HARDY B and HETTEL G P (2002), *Rice Almanac,* 3rd edn, LosBanos, Philippines, International Rice Research Institute.

MARSHALL W E (2004), 'Utilization of rice hull and rice straw as adsorbents', in

Champagne E T (Ed.), *Rice: Chemistry and Technology,* 3rd edn, St. Paul, MN, USA, AACC International, 611–630.

NISHITA K D and BEAN M M (1982), 'Grinding methods: Their impact on rice flour properties', *Cereal Chem,* 59, 46–49.

ORTHOEFER F T and EASTMAN J (2004), 'Rice bran and oil', in Champagne E T (Ed.), *Rice Chemistry and Technology,* 3rd edn, St. Paul, MN, AACC International, 569–593.

PILLAIYAR P(1988), *Rice: Postproduction Manual,* New Delhi, India, Wiley Eastern.

van RUITEN H T L (1985), 'Rice milling: An overview', in Juliano B O (Ed.), *Rice: Chemistry and Technology,* 2nd edn, St. Paul, MN, USA, AACC international, 349–388.

SATAKE T (1990), *Modern Rice-Milling Technology,* Tokyo, Japan, University of Tokyo.

SHARIF M K, BUTT M S, ANJUM F M and KHAN S H (2014), 'Rice bran: A novel functional ingredient', *Crit Rev Food Sci Nutr,* **54**, 807–816.

STEFFE J F, PAUL SINGH R and MILLER G E Jr (1980), 'Harvest, drying and storage of rough rice', in Luh B S (Ed.), *Rice: Production and Utilization,* Westport, Connecticut, Avi, 311–359.

VERGARA B S (1980), *'Rice plant growth and development',* in Luh B S (Ed.), Rice: Production and Utilization, Westport, Connecticut, USA, The Avi Publishing Co. Inc., 75–86.

WADSWORTH J I (1991), 'Milling', in Luh B S (Ed.), *Rice: Vol I, Production,* 2nd edn, New York, Avi Book Published by Van Nostrand Reinhold, 347–388.

WASSERMAN T, FERREL R E, BROWN A H and SMITH G S (1957), 'Commercial drying of Western rice', *Cereal Sci Today,* 2, 251–254

WIKIMEDIA 1 (2014),

http://commons.wikimedia.org./wiki/File:Chinese_rice_mill_1804.jpg

WIKIMEDIA 2 (2014),

http://commons.wikimedia.org./wiki/File:Engelbert_1904_huller_jpg

WIMBERLEY J E (1983), *Technical Handbook for the Paddy Rice Postharvest Industry in Developing Countries,* Los Banos, Philippines, International Rice Research Institute.

WINROCK INTERNATIONAL (1990), *Energy from Rice Residues,* Bioenergy Systems Report, Arlington, VA, USA, Winrock International Institute for Agricultural Development.

YEH A-I (2004), 'Preparation and applications of rice flour', in Champagne E T (Ed.), *Rice: Chemistry and Technology,* 3rd edn, St. Paul, MN, AACC International, 495–539.

Index